A First Course in Laboratory Optics

An optics experiment is the product of intricate planning and imagination, best learned through practice. Bringing forth the creative side of experimental physics through optics, this book introduces its readers to the fundamentals of optical design through eight key experiments. The book includes several topics to support readers preparing to enter industrial or academic research laboratories. Optical sources, model testing and fitting, noise, geometric optics, optical processes such as diffraction, interference, polarization, and optical cavities are just some of the key topics included. Coding tutorials are provided in the book and online to further develop readers' experience with design and experimental analysis. This guide is an invaluable introduction to the creative and explorative world of laboratory optics.

Andri M. Gretarsson is Associate Professor at Embry-Riddle Aeronautical University. A member of the LIGO Collaboration for the detection of gravitational waves, he helped commission the initial detectors and researched the effect of optical coatings on a range of current and future detectors. He has 15 years' experience teaching laboratory optics.

A First Course in Laboratory Optics

ANDRI M. GRETARSSON

Embry-Riddle Aeronautical University

CAMBRIDGE
UNIVERSITY PRESS

CAMBRIDGE
UNIVERSITY PRESS

University Printing House, Cambridge CB2 8BS, United Kingdom

One Liberty Plaza, 20th Floor, New York, NY 10006, USA

477 Williamstown Road, Port Melbourne, VIC 3207, Australia

314–321, 3rd Floor, Plot 3, Splendor Forum, Jasola District Centre, New Delhi – 110025, India

79 Anson Road, #06–04/06, Singapore 079906

Cambridge University Press is part of the University of Cambridge.

It furthers the University's mission by disseminating knowledge in the pursuit of education, learning, and research at the highest international levels of excellence.

www.cambridge.org
Information on this title: www.cambridge.org/9781108488853
DOI: 10.1017/9781108772334

First published 2021

A catalogue record for this publication is available from the British Library.

ISBN 978-1-108-48885-3 Hardback

Additional resources for this publication at www.cambridge.org/gretarsson

Contents

Preface

The aim of this book is to prepare you for work in a field requiring the use of laboratory optics. Through the experiments, you will build and investigate several of the most important fundamental optical systems. While each experiment is preceded by several pages of theory, this book is not intended to be a comprehensive textbook about optics. There are already many excellent such texts. The idea here is to learn optics by *doing optics* with theory as a backdrop. The theory is presented "purposefully" in order to support the experiments and covers the core elements of each topic, enough for you to really understand what's going on. The idea is that after reading this book and doing the experiments, you'll be able to walk into an industrial or academic research lab and at least know how to get started!

The book has one experiment per chapter (except for the introductory first chapter). Each experiment should take about eight hours to complete. You're given the goals of each experiment and just enough information to get started on each part. After that it's up to you to figure out how to proceed based on the theory and other information you may gather. I've tried hard to avoid giving you "cookbook style" experiments so that you have the opportunity to do your own creative experimental design work. As a result, you'll need to give yourself time to investigate different approaches to a problem. Be careful not to get too deep into an experiment before checking whether your approach is yielding sensible data. Before starting an experiment, make sure you study the relevant chapter(s), do some of the problems and make a preliminary design/plan for each section of the experiment. If you can, visit the lab during the design phase and look at the equipment you'll be using. The experiments are all fairly open-ended. Extending them to include follow-on investigations is always possible and may be expected. A few ideas for follow-on work are included at the end of each experiment but you will need to tailor your investigations to the equipment available/obtainable. In addition to the equipment listed in the experiments, you will need access to an optics bench (either a sizable optical breadboard or an optical table), a set of mirrors and lenses, and a selection of mounting hardware for these and other optical components. Most of the experiments rely on using a computer with Matlab, Python, and so on, for analysis tasks. A small selection of Matlab and Python 3 code for planning and analysis tasks is provided in the appendixes. The Matlab code has also been tested in Octave. While the code can be used verbatim, it's primarily intended as a framework upon which to expand. The Matlab and Python versions of the code are available to download at https://github.com/CambridgeUniversityPress/FirstCourseLaboratoryOptics. Additional support materials are also placed there and errata as they arise.

It's worth mentioning the word "intensity." In much of physics, intensity is synonymous with energy flux density. However, in the subfield of optics known as radiometry, "intensity" is sometimes used as a shorthand for "radiant intensity," which has a different meaning. This causes confusion and I try to avoid using the unqualified term "intensity" altogether. I use the term *irradiance* for energy flux density associated with light, that is, "power per unit area perpendicular to the direction of the light's travel." I generally use the letter I for irradiance.

Instructors

I recommend using the book as part of a laboratory-based course in optics with co-requisite lectures. Advanced students will be able to do the labs without formal lectures and the relevant theory sections can be assigned for self-study. In order to reduce the amount of equipment needed for a class, it is helpful to run three or four experiments simultaneously. That way, only two or three setups are required for each experiment. I recommend assigning the relevant theory sections and several exercises as preparation for each experiment. The exercises vary widely in difficulty and some rely on moderate coding ability. Although the number of exercises provided is modest, I hope it is possible to find a mixture that suits the student level and available time.

The text is aimed at upper-level undergraduates and beginning graduate students. The book assumes preparation in vector calculus and at least one physics laboratory course. Some prior programming experience is helpful but an introduction to Matlab and Python is also included in Appendix A. Scientists in other fields embarking on experiments requiring a significant optical component should also find the text useful. By and large, the theory is covered in a survey mode. However, the propagation of laser beams through optics chains is so central to experiments in optics that it is covered more thoroughly. Specialty topics like nonlinear optics, laser cooling, advanced imaging, holography, scattering theory, and so on, are not addressed. The notation and theory development largely follows that of popular optics and laser books, such as *Optics* (Hecht 2017); *Principles of Physical Optics* (Bennett 2008); *Introduction to Modern Optics* (Fowles 1989); *Principles of Lasers* (Svelto 2010); and *Introduction to Optics* (Pedrotti 2007). I've found all of these books to be good supporting texts for the laboratory-based optics class that I teach. As for more advanced texts, I recommend *Lasers* (Siegman 1986); *Photonics: Optical Electronics in Modern Communications* (Yariv 2007); and of course Born and Wolf's *Principles of Optics* (Born and Wolf 2019). There are two other books on laboratory optics I recommend as complementary reading to this text. They are *An Introduction to Practical Laboratory Optics* (James 2014), and *Laboratory Optics: A Practical Guide to Working in an Optics Lab* (Beyersdorf 2014), which is a multimedia book.

Acknowledgments

This book was motivated by a decade of teaching laboratory optics to third- and fourth-year undergraduate students. I'm grateful to the students who came through my course and tried hard to get their experiments to work even when the instructions were impossibly vague. Thanks to them, I have settled on what I hope is "just the right amount" of information to give students who are embarking on an experiment. I owe thanks to the students who pointed out significant manuscript errors and/or gave constructive feedback about the course and/or manuscript content. They include Harsh Menon, Billy Nollet, Aaron and Amy Rose, Marina Koepke, Calley Tinsman, Ashley Elliot, and Brennan Moore. I have surely forgotten some contributions, for which I apologize. Thanks to Professor Darrel Smith, who as the department chair encouraged me to create a serious laboratory optics course for upper-level undergraduates and allowed me a free hand in its design. Thanks to my mother, Anna Garner, for carefully proofreading the manuscript. Special thanks to my wife, Ellie Gretarsson, for testing experiments, reading the manuscript, and giving invaluable feedback at all stages.

1 Light Waves

1.1 Maxwell's Equations

The field of optics describes the behavior of light as it propagates through space and materials. To understand the behavior of light, we start with the fundamental classical physics model describing it: Maxwell's equations of electrodynamics. Maxwell's equations show that the electric and magnetic fields can travel as waves. In a source-free region, Maxwell's equations in linear media are[1]

$$\vec{\nabla} \cdot \vec{E} = 0, \tag{1.1}$$

$$\vec{\nabla} \cdot \vec{B} = 0, \tag{1.2}$$

$$\vec{\nabla} \times \vec{E} = -\frac{\partial \vec{B}}{\partial t}, \tag{1.3}$$

$$\vec{\nabla} \times \vec{B} = \mu\epsilon\frac{\partial \vec{E}}{\partial t}, \tag{1.4}$$

where $\vec{E}$ is the electric field, $\vec{B}$ is the magnetic field, μ is the permeability of the medium, and ϵ is the permittivity of the medium. If we take the curl of both sides of Eq. (1.3), apply the vector identity $\vec{\nabla} \times \vec{\nabla} \times \vec{E} = \vec{\nabla}(\vec{\nabla} \cdot \vec{E}) - \nabla^2\vec{E}$ to the left-hand side and exchange the order of the time derivative and the curl on the right-hand side, we get

$$\vec{\nabla}(\vec{\nabla} \cdot \vec{E}) - \nabla^2\vec{E} = -\frac{\partial(\vec{\nabla} \times \vec{B})}{\partial t}. \tag{1.5}$$

Then substitute from Eqs. (1.1) and (1.4) to get

$$\nabla^2\vec{E} = \mu\epsilon\frac{\partial^2 \vec{E}}{\partial t^2}. \tag{1.6}$$

This is the wave equation in three dimensions where the wave speed is $v = 1/\sqrt{\mu\epsilon}$. Taking the curl of Eq. (1.4) and performing similar algebra shows that the magnetic field also satisfies the wave equation with the same wave speed. Thus Maxwell's equations allow for electromagnetic waves. In vacuum, the speed is $v = 1/\sqrt{\mu_0\epsilon_0} \equiv c$, the speed of light in vacuum. Light is indeed an electromagnetic wave.

We now look for solutions to Eq. (1.6) and its magnetic field counterpart. We actually only need to solve for the electric field because the magnetic field can always be found from $\vec{B} = \frac{1}{c}\hat{k} \times \vec{E}$, where $\hat{k}$ is the direction of travel. (See Exercise 8.1.) We'll assume

---

[1] A source-free region has no net free charge and no net free current. Linear media include vacuum and dielectrics: air, glass, and so on. For an excellent introduction to electrodynamics, see Griffiths (2017).

a single linear polarization and single-frequency (monochromatic) electromagnetic wave. The electric field should then be in the form

$$\vec{E}(\vec{r}, t) = \hat{n}\, E(\vec{r}) \cos[\omega t + \phi(\vec{r})] = \mathrm{Re}\left\{\hat{n}\,\tilde{E}(\vec{r})\,e^{i(\omega t)}\right\}, \tag{1.7}$$

where $\tilde{E}(\vec{r}) = E(\vec{r})e^{i\phi}$. Generalizing to the complex plane, we look for solutions to Eq. (1.6) of the form

$$\vec{E}(\vec{r}, t) = \hat{n}\,\tilde{E}(\vec{r})e^{i\omega t}, \tag{1.8}$$

with the anticipation that at the end we will take the real part to get the actual physical field. Substituting Eq. (1.8) into Eq. (1.6) allows us to eliminate $\hat{n}$, reducing it to a scalar equation. Also, the time derivatives can be performed explicitly bringing down an ω^2 from the $e^{i\omega t}$. This results in a second-order partial differential equation known as the Helmholtz equation

$$(\nabla^2 + k^2)\,\tilde{E}(\vec{r}) = 0, \tag{1.9}$$

where $k = \omega/v$ is the wave number. If we can solve Eq. (1.9) for the complex scalar field $\tilde{E}(\vec{r})$, then we can get the actual physical electric field by multiplying our solution by $\hat{n}e^{i\omega t}$ and taking the real part. Since Eq. (1.9) is a second-order partial differential equation in three spatial coordinates (e.g. x, y, z), we will need to specify appropriate boundary conditions for the field on some surface in order to obtain explicit solutions.

1.2 Huygens' Principle

In many cases of interest in optics, Eq. (1.9) is solved to a *good approximation* by Huygens' integral. The field is assumed to be known on a "source plane" S_1 perpendicular to the z-axis and is only nonzero in some finite region of that plane. The values of the field on S_1 serve as a boundary condition for solving Eq. (1.9). The solution is given by Huygens' integral for the complex scalar field at any desired point x, y, z.

$$\tilde{E}(x, y, z) = \frac{i}{\lambda} \iint\limits_{S_1} \tilde{E}(x', y', z') \cos\phi\, \frac{e^{-ik\imath}}{\imath}\, \mathrm{d}S'. \tag{1.10}$$

The integration over the source plane S_1 is performed using the integration variables x', y', z'. The vector $\vec{\imath}$ joins points in the source plane S_1 with the point (x, y, z) at which we are calculating the field. The angle between $\vec{\imath}$ and the z-axis is ϕ (see Figure 1.1). The solution represented by Huygens' integral is satisfying because it encapsulates an intuitive understanding of how light waves behave that was understood long before the formal mathematics was fully worked out.

The intuitive description of Eq. (1.10) is known as *Huygens' principle*, due to Christiaan Huygens (1629–1695), a Dutch mathematician and scientist. Under Huygens' principle, every point in the source is considered to be emitting light with spherical wavefronts propagating outward – the so-called Huygens' wavelets. These wavefronts are represented by

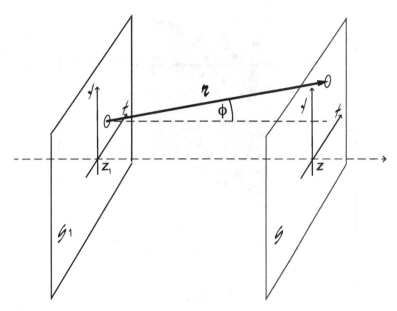

Figure 1.1 The electric field in the source plane S_1 is propagated to the field point in plane S. The source plane is located at $z = z_1$ and the field plane is at z. The circle on the source plane indicates a typical source point involved in the integral. The complex scalar field at this point is $\tilde{E}(x, y, z_1)$. Similarly, the circle on the field plane indicates a typical field point with complex scalar field $\tilde{E}(x, y, z)$.

the factor $\cos\phi \frac{e^{-ik\imath}}{\imath}$. They are emitted preferentially in the direction perpendicular to the source plane due to the presence of $\cos\phi$. The constant $\frac{i}{\lambda}$ out-front contributes 90° of phase and the λ in the denominator serves to keep the units the same on both sides of the equation. The complex scalar field in the source plane $\tilde{E}(x, y, z_1)$ sets the relative amplitudes and phases of these tiny spherical emitters. The field $\tilde{E}(x, y, z)$ is then simply the *linear superposition* of all the spherical wavefronts emitted from the source.

Example 1.1 Single-Slit Diffraction A typical use of the Huygens' integral solution 1.10 is to find the diffraction pattern from a small aperture of some specific shape. Consider the diffraction of a plane wave from a rectangular aperture of width $2b$ and height $2d$ viewed on a screen at a distance L downstream from the slit. The screen distance is much larger than either dimension of the aperture $b, d \ll L$. We choose our optic axis z to be perpendicular to the plane of the aperture and centered on the aperture and choose the x and y axes of the source plane and field plane to be parallel to the width and height of the slit, respectively. The slit is located at $z = 0$ and the screen at $z = L$. The complex scalar field in the slit is assumed to be from a monochromatic plane wave impinging on the slit from the left. See the following diagram. So $\tilde{E}(x, y, 0) = u_0$ for $-b < x \leq b, -d < y \leq d$ and zero otherwise. We want to calculate the complex scalar field on the screen along the x-axis. Taking the squared magnitude of the complex scalar field gives the irradiance.

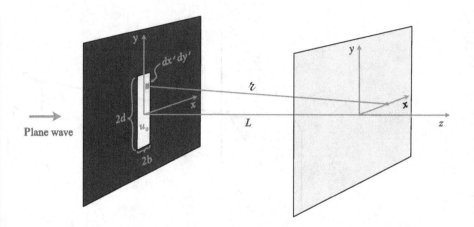

Huygens' principle for this case is then

$$\tilde{E}(x, 0, L) = \frac{i}{\lambda} \int\limits_{x'=-b}^{b} \int\limits_{y'=-d}^{d} u_0 \cos\phi \, \frac{e^{-ik\imath}}{\imath} \, dx' \, dy'. \tag{1.11}$$

For the $\imath$ in the denominator, it's enough to use the approximation $\imath \approx L$. Since $b, d << L$, we also have $\cos\phi \approx 1$. For the $\imath$ in the exponent, where it's multiplied by $k = \frac{2\pi}{\lambda}$ and therefore causes the integrand to vary rapidly, we need the first-order binomial approximation. Since we're going to start by integrating over y', we approximate $\imath$ in terms of $\frac{y'}{L}$ as

$$\imath \approx \sqrt{L^2 + (x - x')^2} + \frac{y'^2}{2L} + \dots . \tag{1.12}$$

The integrand becomes separable and Eq. (1.11) is then

$$\tilde{E}(x, 0, L) \approx \frac{iu_0}{\lambda L} \int\limits_{x'=-b}^{b} e^{-ik\sqrt{L^2+(x-x')^2}} \, dx' \int\limits_{y'=-d}^{d} e^{-ik\frac{y'^2}{2L}} \, dy'. \tag{1.13}$$

The y' integral can be done with the help of tables, computer programs, and so on, but since the integrand depends only on y', the result is just an overall constant, which we shall call A. Then

$$\tilde{E}(x, 0, L) \approx \frac{iA}{\lambda L} \int_{-b}^{b} e^{-ik\sqrt{L^2+(x-x')^2}} \, dx'. \tag{1.14}$$

We can now redraw the previous figure from the viewpoint of a person looking down onto the arrangement from above. The x' integral is now a sum over ribbons of width dx' and having the full height of the slit. This is the way the problem is sometimes represented in introductory physics texts and the issue of why it works (i.e. separability of the integrand) is not explained.

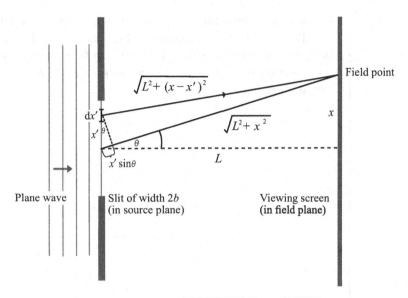

This figure should convince you that $\sqrt{L^2 + (x - x')^2} = \sqrt{L^2 + x^2} - x'\sin\theta$, where θ is the diffraction angle measured from the center of the slit. Making this substitution in Eq. (1.14) and carrying out the integral, we get

$$\tilde{E}(x, 0, L) \approx \frac{2iAbu_0}{\lambda L} \frac{\sin(kb\sin\theta)}{kb\sin\theta} e^{-ik\sqrt{L^2+x^2}}. \tag{1.15}$$

The irradiance I is proportional to the magnitude of the complex scalar field squared:

$$I \propto \tilde{E}(x, 0, L)^* \tilde{E}(x, 0, L) \tag{1.16}$$

$$\propto \frac{sin^2(kb\sin\theta)}{(kb\sin\theta)^2}. \tag{1.17}$$

The maximum irradiance occurs when $\theta = 0$. If the maximum irradiance is I_0, then

$$I(\theta) = I_0 \frac{sin^2(kb\sin\theta)}{(kb\sin\theta)^2}. \tag{1.18}$$

Note that this expression has minima whenever $kb\sin\theta = \pm n\pi$, where $n = \pm 1, \pm 2, \ldots$. Generally, this condition is written in terms of the full width $a \equiv 2b$ of the slit as

$$a \sin\theta_n = n\lambda. \tag{1.19}$$

This equation is familiar to students of introductory optics. If we then plug in $\sin\theta = x/\sqrt{x^2 + L^2}$, we get the irradiance in terms of the screen position x. This is the form we would want for comparing to actual measurements on a flat screen like a digital camera image chip, and so on.

$$I(x, 0, L) = I_0 \left(\frac{\lambda}{\pi a}\right)^2 \frac{x^2 + L^2}{x^2} \sin^2\left(\frac{\pi a}{\lambda} \frac{x}{\sqrt{x^2 + L^2}}\right). \tag{1.20}$$

The corresponding minima locations are

$$x_n = \pm\frac{n\lambda L}{\sqrt{a^2 - n^2\lambda^2}}, \qquad n = 1, 2, \ldots \tag{1.21}$$

The figure here shows the irradiance distribution from Eq. (1.20) and minima locations from Eq. (1.21) for the diffraction of a HeNe laser, $\lambda = 632.8$ nm at a slit seven wavelengths wide. The screen (field plane) is 1 m from the slit, so you can see from the x-axis that the diffraction pattern isn't very wide. It falls off quickly away from the central maximum.

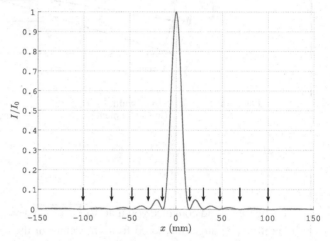

Huygens' principle in the form of Eq. (1.10) can also be used to get a numerical solution using a finite number of discrete emitters. The integral is thus converted to a sum over the fields emanating from these discrete emitters. As an example, Figure 1.2 shows the result of adding the fields propagating from 100,000 identical spherical emitters in a slit 10 wavelengths wide and 100 wavelengths tall. Although the field near the aperture changes quite rapidly, at large distances from the aperture the diffracted field settles into the uniform pattern characteristic of single-slit diffraction.

It is worth re-emphasizing that the value of Huygens' principle lies not only in the fact that we can evaluate the right-hand side analytically or on a computer. It gives us an intuition for the nature of electromagnetic radiation beyond simple plane waves.

1.3 The Paraxial Approximation

We're going to be most interested in the propagation of "beams" of light.[2] Beams of light propagate mostly in one direction and we choose our axes so that the z-axis lies in the direction of propagation. The paraxial approximation is the assumption that all wavefront normals make small angles with the z-axis. It is appropriate for beams and any situation where light travels mostly in one direction. We consider the propagation between two planes perpendicular to the z-axis as shown in Figure 1.1: a source plane $\mathbf{S_1}$ at $z = z_1$

[2] Lasers are the main source of light beams nowadays. For consistency, the approach in this section corresponds closely to that of classic laser textbooks, which should be consulted for extra details. I recommend Svelto (2010) for a more introductory approach and Siegman (1986) for those who want to fill in all the gaps.

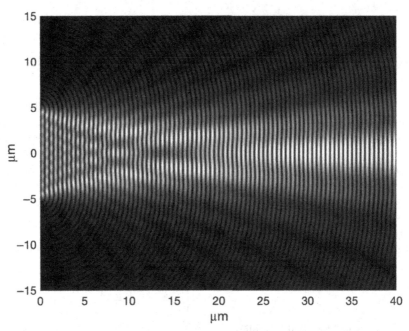

Figure 1.2 Simulation of a plane wave incident from the left, diffracting through an aperture 10 wavelengths across and 100 wavelengths high. The wavelength is 1 μm. The aperture is at the left edge of the image (between −5 and +5 μm). The irradiance was estimated by Huygens' principle using approximately 100,000 spherical Huygens' wavelet emitters in a rectangular grid 10 wavelengths wide and 100 wavelengths high (out of page). The far-field pattern has essentially established itself by the time the light reaches the right-hand edge of the image, 40 μm downstream. A screen placed at the far right of the image, perpendicular to the page, would register the classic single-slit diffraction pattern with the bright central maximum. (The dark and bright bands running mostly vertically are due to the fact that this is a snapshot of the irradiance at a single time. They are the crests, troughs, and zeros of the electromagnetic wave.) To illustrate the fainter features of the diffracted beam, I've plotted the field magnitude (square root of irradiance) rather than the irradiance. This is also closer to the way our eye perceives the irradiance pattern.

and a "downstream" field plane **S** at some unspecified z. As before, we assume that in the source plane, the complex scalar field $\tilde{E}(x, y, z_1)$ is known.

The Helmholtz equation simplifies in the paraxial approximation. Since the light is propagating primarily in the z-direction, it's useful to separate out the rapid phase accumulation in z due to the wave nature of the light by writing

$$\tilde{E}(x, y, z) = u(x, y, z) e^{-ikz}. \tag{1.22}$$

The idea is that as long as the light is traveling largely in the z-direction, the $u(x, y, z)$ will vary very little over distances on the order of a wavelength. In other words, our wave can be treated as something close to a plane wave but with a complex amplitude $u(x, y, z)$ that varies *slowly* with position. $u(x, y, z)$ is sometimes known as the complex field amplitude or just the field amplitude. Substituting Eq. (1.22) into Eq. (1.9), we get

$$\frac{\partial^2 u}{\partial x^2} + \frac{\partial^2 u}{\partial y^2} + \frac{\partial^2 u}{\partial z^2} - 2ik\frac{\partial u}{\partial z} = 0. \tag{1.23}$$

In the common case where the wave occupies only a small region in the source plane and the phasefronts are fairly flat – typical characteristics of what we might call "beams" – then u will vary more slowly in the z-direction than in any other direction, namely

$$\left|\frac{\partial^2 u}{\partial z^2}\right| \ll \left|\frac{\partial^2 u}{\partial x^2}\right| \tag{1.24}$$

$$\left|\frac{\partial^2 u}{\partial z^2}\right| \ll \left|\frac{\partial^2 u}{\partial y^2}\right|. \tag{1.25}$$

Also, the fractional change in the slope $\frac{\partial u}{\partial z}$ should be small over a wavelength λ. That is

$$\left|\frac{\Delta\left(\frac{\partial u}{\partial z}\right)}{\frac{\partial u}{\partial z}}\right| \approx \left|\frac{\frac{\partial^2 u}{\partial z^2}\lambda}{\frac{\partial u}{\partial z}}\right| \ll 1. \tag{1.26}$$

Since $k = \frac{2\pi}{\lambda}$ this implies

$$\left|\frac{\partial^2 u}{\partial z^2}\right| \ll \left|2k\frac{\partial u}{\partial z}\right|. \tag{1.27}$$

So, we can drop $\frac{\partial^2 u}{\partial z^2}$ from Eq. (1.23), leaving

$$\frac{\partial^2 u}{\partial x^2} + \frac{\partial^2 u}{\partial y^2} - 2ik\frac{\partial u}{\partial z} = 0. \tag{1.28}$$

This equation is generally known as the paraxial wave equation.

A solution to the paraxial wave equation can be obtained from Huygens' integral. In the paraxial approximation, where all propagation is close to the optic axis, the pathlength r in Figure 1.1 can be approximated as

$$r = \sqrt{(x - x')^2 + (y - y')^2 + (z - z_1)^2} \tag{1.29}$$

$$\approx L + \frac{(x - x')^2 + (y - y')^2}{2L} + \cdots, \tag{1.30}$$

where, in the second line, L is the propagation distance ($L \equiv z - z_1$). With this, and $\cos\theta \approx 1$, we can rewrite Huygens' integral Eq. (1.10) as

$$u(x, y, z) = \frac{i}{\lambda L} \iint_{\mathbb{R}^2} u(x', y', z_1) e^{-ik\frac{(x-x')^2+(y-y')^2}{2L}} \, dx' \, dy'. \tag{1.31}$$

This integral solution to Eq. (1.28) allows us to handle most systems involving beams and is the easiest way to propagate the complex scalar field between two planes. The paraxial approximation made here amounts to what is also called the "Fresnel approximation." Equation (1.31) is therefore referred to as Huygens' integral in the Fresnel approximation.

1.4 Coherence

Huygens' principle also allows us to discuss one of the ways in which we classify the statistical properties of light. Light sources are often discussed in terms of their coherence, which comes in two types: temporal coherence and spatial coherence. In a temporally coherent emitter, all of the Huygens' wavelets are emitting at the same frequency and the phase of each individual emitter remains fixed for a *long time* (e.g. many nanoseconds for a HeNe laser). This is known as the coherence time. Lasers have high temporal coherence compared to other sources of light. As a result, lasers tend to be very narrow-band emitters, emitting in only a very narrow band of wavelengths around some nominal wavelength. In lasers, the bandwidth of the output light is referred to as the "linewidth." For example, HeNe lasers, which have fairly narrow linewidths, may emit wavelengths in the band $\lambda = 632.816 \pm 0.001$ nm. The coherence length of a HeNe is the distance traveled by the beam in the coherence time. For a HeNe, it's typically a few tens of centimeters but can be tens of meters for carefully designed units.

High temporal coherence does not in itself require that all the Huygens' emitters have the same phase, only that the phase of each individual emitter should vary slowly. Spatial coherence describes the phase relationship between the different Huygens' emitters. In a source with high spatial coherence, all the emitters are in phase with one another, or nearly so. For example, Young's double-slit experiment only yields the expected diffraction pattern when the spatial coherence of the incident light is sufficient that parts of the beam separated by the slit distance have similar phase. Sources with high spatial coherence can be focused to very small spot sizes and can be collimated so that they approximate plane waves. Note that high spatial coherence does not require high temporal coherence even though they usually occur together. As long as the phases of all the emitters stay the same, spatial coherence is preserved whether the overall phase changes are fast or slow, random or not. In Chapter 6, we discuss the properties of etendue and radiance, which are closely related to spatial coherence. Sources with high spatial coherence will have high radiance and low etendue. As a rule, both the temporal and spatial coherence of lasers are the highest of all light sources, which is the main reason they're so useful.

Example 1.2 Partial Spatial Coherence   Consider a light source like certain LEDs with very small emission regions that possesses partial spatial coherence over its beam. If we consider a transverse cross section of such a beam, then adjacent photons in the cross section will usually be in phase with one another but the greater the transverse distance between the photons, the higher the probability they will have random phase with respect to one another. Imagine taking such a light source, covering its aperture with aluminum foil and poking two small pinholes in the foil. Then you've made a version of Young's double-slit experiment. An interference pattern will be formed wherever the two beams emerging from the holes overlap. The following figure shows how partial spatial coherence reduces the fringe contrast. Both panels show a simulation of the interference pattern formed by illuminating two pinholes ten wavelengths apart and viewed on a flat screen 10 cm away. The panel on the left assumes the source has perfect spatial coherence. The fringes have high contrast. The panel on the right assumes that the fields emanating from the two pinholes are only partially coherent due to an imperfect spatial coherence of the source. The fringe contrast is much lower. The fringe contrast can be characterized by a quantity called the "visibility" which is just the difference between the maximum fringe irradiance and the minimum fringe irradiance, divided by the sum. The visibility of the fringes on the left is clearly higher.

Exercises

1.1 Electromagnetic waves are transverse waves. In what sense are they transverse? Does something actually move up and down *in space*?

1.2 In Section 1.1, we took the curl of Faraday's law, Eq. (1.3), to show that the electric field has wavelike solutions in regions free of charge and current (source-free). Now do the same for the magnetic field by taking the curl of Ampere's law with Maxwell's correction, Eq. (1.4).

1.3 Use the source-free version of Gauss' law, Eq. (1.1), and "Gauss' Law for Magnetic Fields," Eq. (1.2), to show that for a monochromatic plane wave, both the electric and magnetic fields are perpendicular to the direction of propagation. *Hint*: Choose the z-axes so that it lies in the direction of propagation, writing $\vec{E} = \vec{E}_0 \cos(kz - \omega t)$ and $\vec{B} = \vec{B}_0 \cos(kz - \omega t)$.

1.4 Apply Faraday's law to a monochromatic plane wave in a linear medium traveling in the $\hat{z}$ direction to show that $\vec{B}_0 = \frac{1}{v}\hat{z} \times \vec{E}_0$, where v is the speed of light in the medium. $\vec{E}_0$ and $\vec{B}_0$ are the vector-amplitudes of the electric and magnetic fields, respectively. Explain why this relationship implies that the electric and magnetic fields are perpendicular to each other. *Hint*: For a monochromatic plane wave traveling in the $\hat{z}$ direction, $\vec{E} = \vec{E}_0 \cos(kz - \omega t)$ and $\vec{B} = \vec{B}_0 \cos(kz - \omega t)$.

1.5 Show that electromagnetic plane waves solve the Helmholtz equation.

1.6 Show that spherical electromagnetic waves $\tilde{E}(r) = \frac{1}{r}e^{-ikr}$ solve the Helmholtz equation. The Laplacian in spherical coordinates is

$$\nabla^2 = \frac{1}{r^2}\frac{\partial}{\partial r}\left(r^2\frac{\partial}{\partial r}\right) + \frac{1}{r^2\sin^2\phi}\left[\frac{\partial^2}{\partial\theta^2} + \frac{\partial}{\partial\phi}\left(\sin\phi\frac{\partial}{\partial\phi}\right)\right].$$

1.7 Consider a monochromatic plane wave traveling in the $+\hat{z}$ direction and incident on a screen with two small holes. (This screen with the holes is the source plane S_1.) We choose our coordinates so that the holes are on the x-axis at $x = \pm d/2$. For a sufficiently small hole, the integral over the source plane in Eq. (1.31) can be simplified by replacing the integral by the integrand times the area of the hole. Use this approach to find the irradiance pattern on the x-axis of a screen (the field plane, S) a distance L away. Assume ϕ is small. *Hint*: Remember that irradiance in the field plane is proportional to $u(x, y, z)^* u(x, y, z)$.

1.8 Fill in the missing algebra or calculus steps in Example 1.1 and turn the detailed derivation in as a carefully written up solution. Check, using integral tables, Mathematica, WolframAlpha, and so on that the constant A which plays a role in setting the overall field amplitude, depends on wavelength, slit height, and distance to the screen according to

$$A \equiv \int_{y'=-d}^{d} e^{-ik\frac{y'^2}{2L}}\, dy' = (1-i)\sqrt{\frac{\lambda L}{2}}\,\mathrm{erf}\left[\sqrt{\frac{2\pi d^2}{\lambda L}}\left(\frac{1}{2} + \frac{i}{2}\right)\right] \approx 2d \quad \text{for } \frac{d}{\lambda} << \frac{L}{d}.$$

1.9 Using Eq. (1.29), verify that Eq. (1.31) follows from Eq. (1.10) as claimed.

1.10 Convert the "several meter" coherence length of a HeNe laser to the equivalent coherence time, t_c, in nanoseconds. Estimate the corresponding linewidth, $\Delta f = \frac{1}{t_c}$, in megahertz. What is the ratio of the linewidth to the laser frequency? So, are typical HeNe frequency variations parts per million, parts per billion, or something else? For comparison, estimate the fractional change in the frequency of a decent mechanical oscillator like a steel tuning fork when its temperature changes by one-hundredth of a degree Celsius.

1.11 (Computer problem) Consider a $\lambda = 633$ nm plane wave traveling in the $+\hat{z}$ direction and incident on a $5\,\mu\mathrm{m} \times 5\,\mu\mathrm{m}$ square aperture. The aperture lies in the xy-plane, centered on the origin. After passing through the aperture and propagating to $z = 1$ m, the light is incident on a screen parallel to the xy-plane. Find the irradiance distribution on the screen by applying Huygens' principle. *Hint*: Express the irradiance distribution on the screen due to each emitter as an $n \times n$ array. In a for-loop, add the irradiance distributions from a square 10×10 grid of Huygens' wavelet emitters (integrand of Eq. [1.2]). Note that $u(x, y, z_1)$ is constant for all emitters but $\imath$ and $\cos \phi$ are different for each emitter. Take $u(x, y, z_1) = 1$ and ignore the prefactor $\frac{i}{\lambda}$ since it only contributes an overall constant.

1.12 (Computer problem) Repeat exercise 1.11, but this time make the aperture circular.

2   Components and Methods

This chapter describes most of the general use optical components you will encounter and their limitations. It also describes the basic methods of analyzing experimental data. The chapter begins with an introduction to laboratory safety.

2.1 Safety

Working safely in an optics laboratory requires you to recognize dangers in situations you probably have little experience with. That requires you to move deliberately. *Never rush.* You must be aware of what you are doing at all times and be careful not to be on "autopilot." As a general rule, don't have casual conversations or allow yourself any distractions while working with or around a laser beam. In your mind, flag areas where hazards may reside and pay extra attention when working in those areas.

2.1.1 Laser Safety

This section is not a replacement for a laser safety course (nor indeed is this book). Completing this section does not qualify you to operate a laser. The designated Laser Safety Officer (LSO) at your institution can qualify you to operate a laser at your institution. The LSO is the individual who has been designated by your institution as the person responsible for overseeing laser safety and evaluating and controlling laser hazards. It is the responsibility of the LSO to ensure that you are properly trained to operate any lasers. The goal of this section is to provide general advice that I hope will encourage you to develop good working habits that will make it easier for you to work safely with and around lasers. Do not use any laser rated above 3R (or 3a, IIIa) for the experiments in this book. In fact, all the experiments can be accomplished perfectly well with Class II lasers.

When Must You Wear Laser Safety Glasses?

The main reason people get an eye injury while working around lasers is that they aren't wearing the appropriate laser safety eyewear (or aren't using it correctly). Lasers are classified according to the level of hazard they present to the user. There are four main laser hazard classes: Class 1, Class 2, Class 3, and Class 4. Class 1 lasers are safe in normal use, while Class 4 lasers are extremely hazardous. Class 3 is split into two subclasses: 3R and

3B (or in older classification systems "3a and 3b" or "IIIa and IIIb"[1]). Lasers of class 3B and class 4 are extremely dangerous to the eyes.

> **You MUST wear the correct laser safety glasses when working with Class 3B or Class 4 lasers.**

If the beam from a class 3B laser or a class 4 laser enters your eye, you are *very likely* to receive an eye injury. Even viewing the spot made when a class 4 laser falls onto some surface can cause injury if you aren't wearing eye protection. In addition, focused class 3B lasers and all class 4 lasers are a skin and fire hazard.

It's also recommended to wear laser safety glasses when working with class 3R (3a, IIIa) lasers, especially during setup. If you get the beam from a class 3R (3a, IIIa) laser into your eye, your blink response may prevent damage but it's also possible that you will suffer an eye injury.

The class of a laser must be printed on the laser housing. It's usually shown somewhere near the aperture. Never use a laser whose class you do not know. If there is no label telling you the class and the wavelength, don't use the laser! Also, never use a laser that has been tampered with because this can affect the output beam in unexpected ways. (Some lasers, such as green laser pointers, are actually frequency-doubled infrared lasers. The infrared component is typically prevented from leaving the laser by a filter. If that filter is broken or removed, you could receive a very hazardous eye exposure to an infrared beam without realizing it.) If the warning label on the laser says words like "danger" or "avoid exposure to beam" take it seriously!

Laser glasses should be obtained from a reputable supplier and should conform to national certification requirements. The ability of laser glasses to block radiation is described by the optical density (OD) rating for the wavelength of the laser in use. Higher OD, means more light attenuation. Note that an OD rating that is sufficiently high to be safe at one wavelength does *not* imply that the glasses are safe for any other laser wavelength. In general, different laser wavelengths require different safety glasses. The OD rating is a logarithmic scale that specifies the transmission of the glasses at the rated wavelength.

$$\mathrm{OD} = \log_{10}\left(\frac{1}{T}\right), \tag{2.1}$$

where T is the transmittance of the glasses. The transmittance of the glasses is the fraction of incident power that is transmitted through the glasses. For example, if a set of laser glasses has an OD of 5 at 1,064 nm, they will transmit 1/100,000th of the incident power at this wavelength.

---

[1] There are three major classification schemes for lasers, all of which you may see used on a warning/danger label: The scheme published by the International Electrotechnical Commission (IEC) is used globally. The IEC laser classes are: 1, 1M, 2, 2M, 3, 3R, 3B, and 4. The Center for Devices and Radiological Health (CDRH) publishes another widely used scheme with classes: I, II, IIIa, IIIb, IV. The American National Standards Institute (ANSI) publishes another scheme with classes: 1, 2, 3a, 3b, and 4. The good news is that the main numbers (1–4) in each scheme are roughly equivalent. In all schemes, any laser marked 3B (or 3b, or IIIb) or 4 (or IV) is very hazardous to the eye.

2.1.2 Good Safety Habits

If you inculcate good habits when working with lasers and associated optics, laser safety will become a natural way of working and not a burden. Many of these habits will also help with lab productivity in general. Here's a list.

Remember the safety glasses

Keep safety glasses in a particular place. Make a habit of putting them on *before* turning on the laser! For keyed lasers, it can be a good reminder to store the key near the glasses.

Keep upright

Never put your eyes at the height of the beam. Other than wearing your safety glasses, this is the most important precaution you can take. Similarly, don't bend down to get a dropped object or lean down over the optic chain in such a way as to bring your eyes near the height of the laser. When you are not upright, your safety glasses may not sit properly, potentially exposing you to a very hazardous situation.

No jewelry

Remove all reflective objects on your body or clothes before working with a laser. This includes watches, smartwatches, rings of all types, brooches, and so on. Hair should be tied back so as not to fall into the beam and so as not to contaminate the optics.

Horizontal beam path

If at all possible, keep the beam path at a single uniform height. Most importantly, avoid any beams that don't travel horizontally. If you must change the beam height, use periscopes specifically designed for that purpose.

Close the shutter first

Always block the laser before adding or removing an optic. Usually, this is as simple as closing the laser shutter. Align a new optic as best you can *without* the beam. Then unblock the laser and adjust the alignment as required.

Block stray beams

Anticipate stray beams and place a beam block in their path. For invisible and higher power lasers, use beam blocks specifically intended for the wavelengths and powers in use. For low-power visible lasers, stray reflections can often be dumped on the mount of the preceding optic but don't let them leave your table. Before turning off the laser, make a habit of scanning for stray beams. This is especially important if you are working with infrared or other invisible beams. Once you are sure there are no stray beams, the first step in turning off the laser is to close the laser shutter.

Take a break

It's dangerous to work tired, hungry, rushed, or distracted. Take a break instead and come back to it later. Even if you don't get injured, you are likely to break something or make a poor decision that will set you back. Believe me, taking a break is not only the safe choice, it's also the most productive.

Don't look into the laser aperture

This may seem straightforward and most people are sure they would never do this. However, it's a natural temptation to look into a laser to see why it's not turning on. Most laser power supplies have a couple of seconds delay between keying the switch and actual turn on. This interval is long enough to tempt some people to look into the aperture to see what's going on. Don't do this! Similarly, laser pointers are all too easy to flip around to look at the aperture. Don't! The correct approach in each case is to get a small card and place it in front of the aperture. It will reveal the presence or absence of the beam quite nicely. In general, beams can be conveniently located through the use of a card like this. For invisible wavelengths or any wavelengths fully blocked by safety glasses, special beam-viewing cards are used that light up at visible frequencies in response to the beam.

2.1.3 Non-Beam Hazards

Non-beam hazards are often overlooked. Probably the most common non-beam hazard is electric. Many lasers have power supplies that operate at high voltage. In some cases, these voltages may be exposed or otherwise accessible to the user. For example, HeNe lasers require several thousand volts to operate, especially during turn on. Commercial power supplies for *low power* HeNes (< 5 mW) are typically limited to pretty low current (a few milliamps) and aren't likely to kill someone. However, power supplies for larger HeNes or home-built/modified power supplies can be quite lethal indeed. Power supplies for most lasers are dangerous. Don't mess around with laser power supplies and don't try to repair or modify them.

The dyes used in dye lasers can be very toxic. The solvents used with these dyes may also be toxic, carcinogenic, and/or flammable.

If you are working with a high-power or short wavelength laser, the beam may interact with a target to produce toxic or irritant byproducts. Lasers used for cutting or etching can generate significant amounts of such by-products and need to be vented. Make sure the ventilation system is on and working. This is far from a comprehensive list, so make sure you are familiar with any non-beam hazards associated with your use of any particular laser.

Optics work is easier and safer if you don't have to worry about snagging your clothing on optical mounts, and so on. Wearing clothing that's unlikely to snag is a good idea. Lab coats can help with this. Make sure you have either close-fitting sleeves and cuffs, short or elbow-length sleeves, or just roll your sleeves up out of the way. General laboratory safety advice also includes wearing closed-toe shoes while working.

2.2 Optical Components

This section introduces you to common optical components in order to give you some idea of how to choose components for your experiments.

2.2.1 Light Sources

Practical light sources fall into two main categories: thermal sources and quantum level-transition sources. Thermal light sources are just hot objects, like an incandescent light bulb filament (hot tungsten), a flame (hot gases), or a star (hot hydrogen). They emit very broad-spectrum light unless they are filtered in some way. Level-transition light sources on the other hand emit due to a transition from a higher energy level to a lower one, the energy difference being emitted as a photon. Typical examples are: excited states of gases in discharge lamps or gas lasers; level transitions across a semiconductor junction as in an LED or a diode laser. Such light sources tend to be narrow band or contain a set of narrow emission bands. There are some sources of "hybrid" light such as high-voltage arc lamps that generate an underlying thermal spectrum with some very broad emission lines in the spectrum as well. Another example of such a hybrid source is the standard fluorescent lamp. The tubes themselves are mercury vapor discharge tubes but the inside surface of the tube has been coated with a special coating that fluoresces in the visible when excited by UV light from the mercury emission; hence their name. In the following subsections, I discuss some of the common laboratory light sources you can expect to use.

Incandescent Lamps

The brightest incandescent lamps are so-called halogen lamps (Figure 2.1). Laboratory-grade halogen lamps are also called "QTH" lamps. This stands for quartz-tungsten-halogen and refers to the quartz glass envelope (the bulb itself), the tungsten filament, and the halogen added to the otherwise inert fill-gas within the bulb. These lamps are quite similar to the old incandescent household light bulbs or low-power automotive bulbs where a filament sitting in an inert gas is heated by passing a current through it, until it glows brightly. The main difference between a halogen lamp and a standard incandescent lamp is that the halogen lamp generally runs hotter and therefore gives off a brighter white light. The halogen gas prolongs what would otherwise be a short filament life by depositing evaporated tungsten back onto the filament during operation. The underlying emission spectrum is that of a gray body. A gray body emits a spectrum of the same shape as a blackbody at the same temperature but a gray body does not emit as brightly. However, the spectrum may be altered somewhat by the glass envelope, any intentional coatings, and by the fill-gas. The envelope is often made of fused quartz – a glass that may transmit ultraviolet light from the filament (depending on how its coated). For many uses, a laboratory-grade QTH lamp may be unnecessary. An automotive-style halogen bulb may suffice, or even a standard incandescent bulb.

Incandescent lamps have very low spatial and temporal coherence and so can't be used for tasks that require one part of the beam to have a fixed phase with respect to another part. However, some incandescent light sources may have very good amplitude noise properties.[2]

---

[2] For example, the humble flashlight with a traditional incandescent bulb is shot noise limited "out of the box."

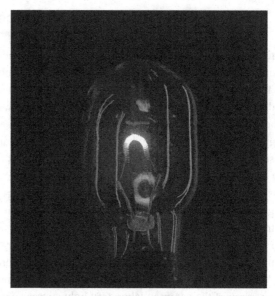

Figure 2.1 A halogen bulb showing the quartz envelope and filament. The fill-gas is transparent and therefore invisible. The lamp is normally too bright to look at directly. (For this picture, the current was greatly reduced).

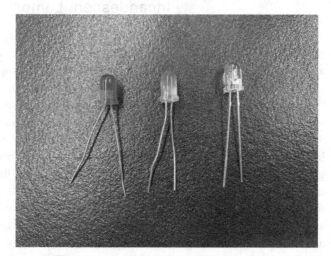

Figure 2.2 Examples of LED lamps. Some LEDs have tinted plastic envelopes so that the LEDs color is evident even when the LED is off. The longer lead indicates the positive terminal. LEDs are simplest to operate from a several volt DC source passed through a several kiloOhm resistor. The effect is to limit the current to a few milliamps, above which the LED will burn out.

LEDs

Light Emitting Diodes (Figure 2.2) can be considered an intermediate step between an incandescent lamp and a laser. LEDs are more coherent than incandescent lamps but less

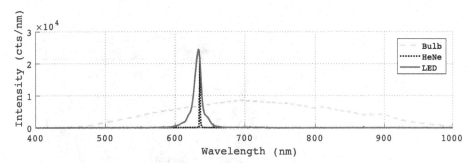

Figure 2.3 The spectrum of a red LED compared with the spectrum of a halogen bulb and a laser. The LED spectrum is fairly narrow band. The actual width of the HeNe line is narrower than can be registered by this spectrometer. Conversely, the bulb's response is too wide to be well represented by this raw spectrum. The bulb's emission does fall off toward shorter wavelengths as shown but actually has rising emission in the infrared; the spectrometer sensitivity falls off toward longer wavelengths leading to the response seen here.

coherent than lasers. Figure 2.3 compares the spectrum of an LED to that of a laser and a halogen bulb. The color of lightemitted by an LED depends on the material used to fabricate the actual semiconductor junction and is not controlled by the color of plastic encapsulating the light-emitting region. A "clear" LED may emit any color light. Generally, LEDs with smaller light-emitting regions, like micro-LEDs, are most spatially coherent while LEDs with narrow bandwidths, like single color LEDs, are most temporally coherent. LEDs are not usually considered sufficiently coherent to do interferometry but may form low-contrast diffraction patterns when passed through a double slit or similar.

Lasers

Lasers are such a broad class of light source that I'll only describe the two types that you are most likely to come into contact with in a first optics laboratory and indeed in everyday life: The helium-neon (HeNe) laser and the diode laser.

The HeNe is a very capable laser with a narrow linewidth (from megahertz to gigahertz depending on the unit). Some HeNe lasers have excellent amplitude noise properties and may even be close to shot noise limited. Shot noise is white noise generated by the "pitter-patter" of photon wavefunctions collapsing at a photodetector. Meanwhile, other HeNes are very noisy or have "squeals and chirps" running through the audio band at random intervals. Due to their long coherence time, pretty much all HeNes make a good choice for simple interferometry experiments. The polarization properties of HeNes also vary quite a lot. Some have a single polarization while others are randomly or partially polarized. I've seen ill-behaved, partially polarized HeNes where the dominant polarization axis rotates slowly and randomly with time. The upshot is that while HeNes are great, some care is required in order to pick a suitable one for your use. It's also worth mentioning that although most HeNes lase on the 632.8 nm line, it's possible to purchase HeNes that lase at any of

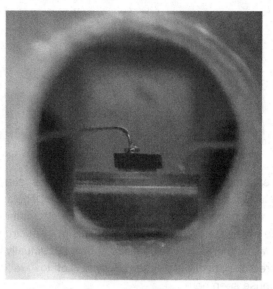

Figure 2.4 If one purchases a bare laser diode (as opposed to a packaged diode laser) one generally receives a small brass "can" about the size of a pencil eraser that has three leads sticking out the back. One of the leads goes to the laser diode itself, the other to a photodiode that is usually incorporated in the can so that the output of the laser can be monitored and controlled. The third lead is the common return (ground) connected to the can. This photo is taken looking into the round aperture (visible in the foreground) where the laser beam comes out. The aperture is about 1 millimeter across. The laser diode is the small black parallelogram in the center of the aperture that is formed as an artificially grown crystal. The formulation of the crystal sets the output wavelength. The electrical contact coming in from above provides the laser current. The other terminal is the metal block on the bottom, which also acts as a heat sink.

several lines in the infrared (1.15, 1.52, and 3.39 μm) or in the visible from green through red (543 nm, 594 nm, 604 nm, 612 nm, and 633 nm).

Laser diodes can be obtained at almost any wavelength and with a huge range of output powers from microWatts to Watts. They're small, cheap, and very efficient at converting electricity into light. As a result, they find uses in everything from supermarket checkouts and laser pointers to advanced laboratory instrumentation. The light-emitting region of a diode laser is quite small and works on similar principles as an LED, except that an optical standing wave is set up in the diode by the means of reflective coatings on opposite ends of a small block of the diode material. The diode itself is often no bigger than a grain of sand. See Figure 2.4. The linewidth of a laser diode is quite large compared to that of a helium neon laser and the coherence time is commensurably short. This makes them less well suited for interferometry unless the optical paths of all beams are very similar. On the other hand, the amplitude of diode lasers can be controlled very easily by controlling the supply current, so very low amplitude noise can be achieved for the cost of a printed circuit board and some modest electronic components. Since the amplitude can also be varied very rapidly, laser diodes can be used in radio frequency signaling, including fiber optic telecommunications. A significant drawback of laser diodes is that they are very easy to destroy through any spikes in the supply voltage. Like LEDs, laser diodes have a nonlinear relationship between the applied voltage and current drawn. The amount of current drawn can also vary as a function of diode temperature. Therefore, they should be operated from a current-limited source. If constant optical power is required, a photodiode can be used to monitor the output. A feedback circuit corrects the supply current to compensate for any changes.

2.2.2 Lenses

There are three basic lens types: plano- , bi- , and meniscus. The profiles of the positive focal length version of each lens is shown in Figure 2.5. Each lens type behaves basically the same way and often it doesn't much matter which type one uses. Common lens kits contain both plano- and bi-type lenses but less often the meniscus type. The difference between the lens types is all about minimizing aberrations. Most lenses are ground with a spherical surface. This is cost-effective[3] but induces small aberrations on the light. Images may not be as sharp and focused beams not as small as they would be if the lenses were diffraction limited (the fundamental limit). Careful use of the right kind of spherical lens can minimize the effect of such aberrations while retaining the cost-effectiveness of spherical surface figures.

Plano- lenses should be used when the rays on one side of the lens are somewhat collimated and the rays on the other side of the lens are converging or diverging. In that case, *the curved part of the lens should face the collimated side.* For example, the smallest and least distorted focus is formed when parallel rays are focused by a plano-convex lens with the curved side of the lens facing away from the focus. When the rate of convergence or divergence is about the same on either side of the lens a bi-lens is preferable.

Meniscus lenses are primarily used in compound lens systems, which are mostly beyond the scope of this discussion. Compound lens systems, such as in cameras and telescopes, can reduce the aberrations over a large field of view. Compound lenses can also reduce chromatic aberration. Chromatic aberration occurs because the index of glasses varies slightly with wavelength. This effect is called dispersion, and causes the effective focal length of a lens to change with wavelength. It's possible to compensate for this

Figure 2.5   Cross sections illustrating the three lens types. From left to right: Plano-convex, bi-convex, positive meniscus.

[3] Lenses with aspherical surface figures tend to be several times more expensive than the equivalent spherical lens.

with a so-called achromatic doublet, where two lenses with opposite dispersion properties are combined.

2.2.3 Mirrors, Reflective Prisms, and Non-Polarizing Beamsplitters

Mirrors, reflective prisms, and non-polarizing beamsplitters are used to change the direction of a beam, flip its orientation, and/or to separate it into two or more similar beams. In general, the shape and propagation characteristics of the beam (rate of divergence, wavefront shape, etc.) are not affected. An exception is the curved mirror which, in addition to redirecting the beam, changes its shape.

A mirror is just a reflective coating deposited onto a substrate, usually glass. The reflective coating is either metallic or of the multilayer dielectric type[4] where constructive interference between many partial reflections from individual coating layers can add to produce very high reflectivity. (See Section 5.1.2.) Laboratory mirrors are usually "first-surface" mirrors, meaning that the reflective coating is right on top of the substrate. That way, a beam doesn't need to pass through a thick layer of glass before reflecting. Some first-surface mirrors are truly bare, meaning that the coating is exposed to air. Such mirrors are very sensitive to handling and fingerprints are ruinous to bare metallic coatings. Most first-surface mirrors actually have a very thin layer of glass deposited over the coating itself. That means they can be cleaned and handled without as much danger of permanent damage. Even protected coatings are delicate, so care with handling is always in order. In all cases, be careful not to touch the coated side. Figure 2.6 shows a selection of mirrors.

Figure 2.6 Various first-surface mirrors. **Left:** aluminum, **Top:** silver, **Bottom:** gold, **Right:** multilayer dielectric.

---

[4] The "dielectrics" in this context are usually transparent glassy materials that are deposited into a multilayer stack such that adjacent layers have different indices of refraction.

The choice of mirror type is entirely dependent on the use. For example, multilayer coatings can reach extremely high reflectance but generally operate effectively only in a narrow band of frequencies. Such mirrors are often called "laser line mirrors" since they are generally designed for particular laser wavelengths. Laser-line mirrors transmit most wavelengths, so they look quite transparent to the naked eye, or they may have a slight hue depending on the wavelength for which they are designed. Since it can be hard to spot such a coating if both sides of the optic are polished, manufacturers will often put a small arrow on the barrel of the mirror, either printed or simply written in pencil, indicating the coated side.

To obtain a multilayer dielectric coating that operates at a broad range of frequencies or at a broad range of incident angles requires compromising on reflectivity to some extent. Therefore, metallic coatings tend to cover the widest wavelength ranges and incident angles while maintaining good reflectivity. (It's also possible to buy "enhanced" metallic coatings on top of which a dielectric coating has been deposited. The dielectric coating changes the reflection properties of the metal while retaining the metal's general broadband reflectance characteristics.) The most common metallic coatings are: aluminum, silver, and gold. Aluminum's reflectance extends into the ultraviolet but is a little lower than silver in the visible. Silver has the highest reflectance at shorter visible wavelengths while gold has the highest reflectance in red and infrared. At very long wavelengths, beyond several microns, multilayer dielectric coatings are hard to obtain as many dielectrics become absorbing at such long wavelengths. So metallic coatings are used in the far infrared. Also, if the phase change upon reflection must be as wavelength-independent as possible, bare metallic coatings are a good choice.

There are several types of reflective prisms allowing you to turn a beam through 45°, 90°, and so on, while at the same time flipping the orientation (parity) of an image about one or both transverse axes. They depend on total internal reflection to do their job but are also sometimes coated. Unlike familiar equilateral prisms, these prisms are not intended to diffract light. They act like mirrors with specific parity properties. Rhomboid prisms only translate a beam but don't change its direction of travel or orientation/parity. Dove prisms only change the orientation/parity but don't translate it or change its direction of travel. An anamorphic prism pair can be used to circularize a diode laser beam that comes out of the laser elliptical in cross section. Trihedral prisms are used to retro-reflect a beam back along the way it came, regardless of the incident angle. (Trihedral prisms are also called retro-reflectors or corner cube reflectors.) Prism types and uses are too numerous to go into more detail here but they are interesting and beautiful objects, and browsing an optics warehouse's prism assortment can be interesting.

A non-polarizing beamsplitter is used whenever you need two similar beams. Don't confuse non-polarizing beamsplitters with their polarizing cousins or you'll get some very unexpected results! Beamsplitters are often used to pick off a small amount of light, such as for a camera, or for amplitude monitoring. "50–50 beamsplitters," which split the light equally are the crucial elements in many optical interferometers. (See Chapter 5.1.) Beamsplitters operate by means of a partially transmissive coating placed on a glass surface that is intended to sit at a particular angle with respect to the incoming light, usually 45°. That way, some of the light is reflected out of the incident beam and some is transmitted through

Figure 2.7   Cube style and plate style beamsplitters.

the beamsplitter. Sometimes the coating is a very thin metal coating – a "half-silvered" mirror – but usually it's a multilayer dielectric coating intended for operation at some specific range of wavelengths. Figure 2.7 shows two different beamsplitter styles, a cube and a plate. Cube beamsplitters generally have better performance, but plate beamsplitters can be made larger and are generally less expensive. In a cube beamsplitter two right-angle prisms are glued together with optical cement. The diagonal side of one of the prisms is coated beforehand. Often you will see a dot, either printed or in pencil, on the ground-glass top surface of the beamsplitter, near one of the optical faces. This indicates which half is supposed to accept the input beam. Making sure the dot faces "upstream" helps to protect the optical cement and ensures that the beamsplitter performs as designed.

2.2.4 Polarizers, Wave Plates, and Optical Isolators

Polarizers come in a bewildering variety and price ranges, from low cost polarizing plastic sheet ("Polaroid") to extremely high extinction ratio crystal polarizers. The extinction ratio is the ratio of the irradiance of the desired polarization component to the undesired one. This is the same as the ratio of the intensities transmitted through a pair of parallel polarizers to the irradiance transmitted through the same set of polarizers after the second one has been rotated by 90°. Polarizers with extinction ratios of 100,000 and higher are commercially available.

Polarizers fall into two broad categories: Polarizers with two outputs where both polarization states are available downstream of the polarizer, and polarizers which absorb one of the polarizations while transmitting only the desired polarization. Most polarizers are "linear polarizers" in that they have linearly polarized outputs. (See Chapter 8.) "Circular polarizers" also exist whose output is circularly polarized light split into right-handed and left-handed circular polarization components. The main thing to remember when choosing a polarizer is that it should work well at your wavelength(s) and have the required extinction ratio. Higher extinction ratio generally means higher cost, so understanding your actual needs is wise. Optical power absorption can be a consideration when operating at high optical power and wavefront distortion may be a consideration when doing interferometry.

The effect of a waveplate is to change the polarization state of the light in a controlled way. It operates on the principle of birefringence: that the speed of light through the waveplate material depends on the polarization state of the light. For example, the vertical polarization component of a beam may travel faster through the waveplate than the horizontal component. The vertical component will therefore get a phase lead compared to the horizontal component. If the phase lead accumulated is 90° (a quarter wavelength) and the two polarization components have equal amplitude then circular polarization results. In general, a waveplate turns linearly polarized light into some form of elliptically polarized light. Waveplates are usually sold mounted and with the "fast axis" – the axis along which a beam must be polarized to travel most rapidly through the plate – marked in some way. The simplest type of waveplate is a sheet of something like crystalline quartz with the crystal axis chosen so that the speed of one polarization is higher than that of the orthogonal polarization. The thickness of the plate is chosen so that after passing through the waveplate, a particular wavelength of light will show the desired phase difference between the polarization states *modulo* an integer number of wavelengths: $\Delta\phi = 2n\pi + \theta$ where θ is the nominal phase difference due to the waveplate. Quarter-wave plates, $\theta = \pi/2$ and half-wave plates, $\theta = \pi$ are most common. Quarter-wave plates are used to turn linear polarization into circular polarization and vice versa. Half-wave plates are used to change the angle of a linear polarization component by reflecting it about the fast axis of the waveplate.

So-called zero-order waveplates, as opposed to the multi-order type described earlier, are made so that the phase difference between the polarizations is precisely $\Delta\phi = \theta$. This is achieved by using two plates where the effect of one compensates for the integer wavelength part of the phase difference. This makes the waveplate much less sensitive to small wavelength changes in the input light. One can also purchase achromatic waveplates that are compensated in such a way as to make the phase difference between the components approximately constant across a range of wavelengths.

In addition to polarizers and waveplates, there are devices that work like one-way gates or "diodes" for light. Such optical isolators are generally used to prevent reflected beams from traveling back up the optics chain and reentering the laser system. Faraday isolators are the most common form of optical isolator. Faraday isolators work by rotating the polarization of light that travels in the wrong direction through the isolator so that it is nullified by a beamsplitter at the upstream entrance to the Faraday. Reentrant light can make a laser noisy, so optical isolators are often placed immediately after the laser. Optical isolators can also prevent unwanted interference due to partial reflections at optical surfaces, sometimes called parasitic interference (see Section 9.3.1).

2.2.5 Electro-Optic and Acousto-Optic Modulators

Optical modulators are used to modulate the amplitude, phase, or frequency of a beam. Electro-optic modulators (EOMs) are based on the fact that some materials, like lithium-niobate, have a birefringence that can be varied by applying an electric field. Figure 2.8

Figure 2.8   A phase modulator. The phase deviation is proportional to the voltage applied via the attached cable. The light enters through the small hole at the lower right of the brass enclosure, passes through the electro-optic crystal, and exits through a similar hole on the far side.

shows a typical EOM. EOMs can be used for imparting a signal onto a beam of light by phase modulation or amplitude modulation. Acousto-optic modulators (AOMs) are based on setting up a sort of "grating" in a material by the use of an acoustic standing wave. The frequency of the standing wave is then added to the frequency of the light passing through the grating. AOMs can be used to shift the frequency of a beam slightly. Both EOMs and AOMs can be used to reduce noise. For example, amplitude-noise on a beam can be reduced by counteracting the fluctuations with an amplitude modulator. The same goes for frequency and phase noise reduction. Provided one has some sort of low-noise reference against which to measure the fluctuations that must be canceled, electro-optic components can be used to produce very low-noise laser beams for applications from atomic and molecular optics to gravitational wave detection.

2.2.6 Cameras and Photodetectors

The last stage in most optical chains consists of some sort of light detection. This could be the eye, but nowadays it's usually a photodiode or other single-element photodetector, or the image sensor in a digital camera.

Figure 2.9   A monochrome CMOS camera with the image sensor exposed.

Cameras in optics chains are often used without attached lenses so that the charge-coupled device (CCD) or complementary metal-oxide semiconductor (CMOS) image sensor is exposed. CMOS has been catching up with the more mature CCD technology and CMOS cameras tend to be less expensive than CCD cameras. When a well-calibrated, low-light image is a critical component of an experiment, such as in astronomical observations, CCD cameras are usually preferred. Whether you use a camera based on CCD technology or on CMOS technology won't matter for the applications we consider. However, the overall quality of the camera will matter. Unfortunately, inexpensive "web cam" type cameras don't suffice to accurately record the irradiance profile of a beam or to record a rapidly changing interference pattern. Any camera used should allow the user to change the settings associated with capturing an image (exposure, amplification, frame-rate etc.,) usually via a computer directly connected to the camera but possibly on the camera itself. Unfiltered silicon-based CCD or CMOS cameras will be sensitive throughout the visible range and out to about 1,000 nm in the infrared. Near-infrared enhanced models are available with better infrared sensitivity extending to about 1, 100 nm. Cameras based on other sensor technologies are available for longer wavelengths but are typically much more expensive than silicon-based cameras. For scientific work, a monochrome camera is generally preferable and the desired wavelength sensitivity can be selected by the use of an external filter. Figure 2.9 shows a typical camera used in an optics lab. The rectangular image sensor is clearly visible with metal traces radiating from it on all sides.

Three types of single-element photodetectors are most common: thermal detectors, photoresistors, and photodiodes (see Figure 2.10). Thermal detectors have an optically absorptive film in front that heats up due to incoming radiation. The resulting temperature

Figure 2.10 From left to right: A thermopile detector (in foreground), a photoresistor and two photodiodes side-by-side in the same can. The dual photodiode arrangement shown here (often called a "split photodiode") can be used to measure small motions of a laser beam. The difference between the photocurrents from the two sides is divided by the sum of the photocurrents. This yields a signal that is insensitive to power fluctuations in the laser.

change is monitored to estimate the incoming power. There are a few different variants depending on how the temperature is measured. A thermopile detector is a thermal detector that uses a special arrangement of thermocouples to measure the temperature change. A bolometer uses thermistors instead of thermocouples for measuring the temperature change. The response of these detectors is quite slow (tenths of seconds) but their sensitivity is extremely broadband. Thermal detectors have essentially no intrinsic wavelength dependence other than the spectral transmission characteristics of the window material in the enclosure.

The photoresistor is an electrical resistor whose resistance decreases as the incident light increases. The response is quite slow (many milliseconds) so they are not suitable for use as high frequency sensors. The advantage of photoresistors is primarily the simplicity of integrating them into a circuit since they can simply replace a normal resistor. They are also very inexpensive.

For the combination of low light performance, large dynamic range, linearity, and versatility, the photodiode is unbeatable. Photodiodes can be biased with a voltage to make them extremely sensitive to low light. They can be made physically small in order to reduce their capacitance, allowing them to be used at radio frequencies. They have enormous dynamic range allowing them to be used without saturation over a very large range of incident power: a single photodiode can be used for optical powers ranging over something like seven orders of magnitude. Typical photodiodes have minimum detectable photocurrents corresponding to only millions of photons per second (assuming 1 sec. integration time). This limit is set by the dark current (the "photocurrent" registered when no light is incident on the detector), which is typically very low, in the pico-amp range. Specialty photodiodes, known as avalanche photodiodes, can detect photocurrents consisting of only

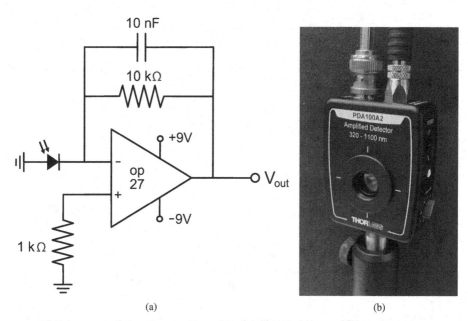

(a)   (b)

Figure 2.11 **(a)** An easy-to-build photodiode amplifier circuit. The $\pm 9\,V_{DC}$ power to the op-amp can be provided by two $9\,V$ batteries placed in series. The ground is taken from the connection between the batteries. The small capacitor is there to prevent oscillation. Higher values do a better job of removing oscillations but at the cost of poor high-frequency response. The $10\,k\Omega$ feedback resistor can be made variable at the cost of slightly higher noise. For best low noise performance, enclose the circuit in a metal box and use BNC bulkhead connectors to bring signals in and out. **(b)** A commercial photodiode and amplifier combination. Commercial units are a convenient alternative and may involve more sophisticated circuits with features like selectable gain, different amplification paths for AC and DC components of the signal, filtering, and so forth.

a few photons, in some cases even just a single photon. The main disadvantage of using photodiodes is that they must be coupled to a very low impedance input. In other words, the device receiving the photocurrent should present little to no impedance to the photodiode. A typical mistake is to plug the photodiode directly into a voltmeter or oscilloscope, both of which present a high impedance to the photodiode. These devices will register the presence of photocurrent but the signal will not be proportional to the light power. Don't do this! Instead, obtain or build a photodiode amplifier. It can be as simple as the circuit shown in Figure 2.11a.[5] For all photodetectors, it's important to know the frequency response. Ideally, you want a flat frequency response over the signal band you're expecting to measure. Photodiodes have the largest bandwidth of all photodetectors but their bandwidth limit depends largely on the capacitance of the photodiode itself. (Sometimes photodiode amplifiers are intentionally given added capacitance to prevent the output from oscillating under

---

[5] The underlying circuit was taken from the classic book on electronics "The Art of Electronics" by Horowitz and Hill (Horowitz (2015)). Every experimentalist should have access to a copy.

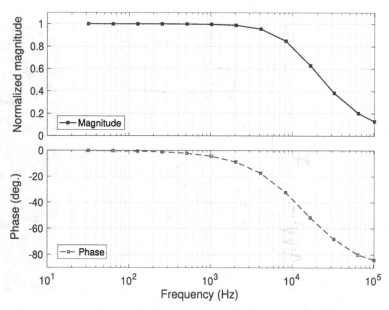

Figure 2.12 The response of a stabilized, large area (~1 cm^2) photodiode amplifier. The magnitude response tends to zero as the frequency tends to infinity. The phase shift tends to $-90°$.

all conceivable conditions reducing the bandwidth further.) Since the intrinsic capacitance scales with the size of the active area of the photodiode, larger photodiodes have intrinsically low bandwidths while very small photodiodes can have high bandwidths. Some photodiode/amplifier combinations have responses extending well into the GHz regime. Figure 2.12 shows an example of the bandwidth of a large photodiode using the photodiode amplifier circuit shown in Figure 2.11a. In this case, the amplifier is intended mainly for low-frequency measurements and the feedback capacitor's capacitance has been chosen quite high in order to provide good stability. The resulting response is limited to below about 25 kHz.

2.2.7 Optomechanics

In the context of an optics laboratory, the word "optomechanics" refers to the hardware used to affix optics into their correct relative position and provide any mechanical motion that may be needed in the course of an experiment. Optomechanics are largely standardized so that parts from one manufacturer can usually be used interchangeably with parts from another. The range of optomechanical components available is enormous but there is a core set of parts just about everyone uses: Optical breadboards/tables with a standard grid pattern of threaded holes, posts to raise the optics to the beam height, and optic mounts to be placed on top of those posts. Figure 2.13 shows a single post set.[6] The components

---

[6] There is another popular system not shown here. It is called a pedestal system. In that case the base, post-holder, and post are replaced by a single pedestal attached to the table with a clamping fork.

Figure 2.13 Work area showing a single optical component and associated optomechanics. *From bottom to top*: Base plate, post-holder, post, optic mount, optic.The black rubber foam mat is there to prevent any dropped components from being damaged and to catch any dropped bolts that could otherwise roll into the holes in the optical table (seen in the background). The lint-free tissue paper will be placed under the immediate work area when the mirror is being mounted into the kinematic optic mount. That way, if it is dropped it won't require cleaning.

are laid out on a soft work surface so that if the optic is dropped during assembly, it will not shatter or chip. Gloves are always used to handle an optic. Often gloves will also be used when handling optomechanical components, especially if the optics are going into a vacuum system or some other clean environment. To assemble the set shown, begin by screwing the base onto the post-holder using the short bolt shown. Make sure it is the correct length so that the threads don't protrude on the inside and reduce the range of motion of the post. After the base is assembled, it should be placed in the approximate correct position on the optical table using the pair of washers and screws shown. If you're working with a laser, it should be turned off or the laser aperture closed. Usually, the optic's position needs to be adjusted transverse to the beam rather than along the beam path. So, the orientation of the base should be such that the slots lie *perpendicular* to the beam path. Note, that in order to protect the base and to allow the slot to move smoothly, the washers should be placed *rounded side down*. The next step is to attach the post to the optic mount, again using the correct length bolt that mustn't be allowed to bottom out in the blind hole at the top of the post. Make sure this bolt is tight (but not brutishly so) or else the optic mount will twist loose when you adjust the post orientation during alignment of the optics chain. Now place the optic post into the post-holder and check that the optic will be at the correct height and position. If so, bring it back to the work area and install the actual optic into the mount. Then immediately place it into the optics chain. Align it as best you can with the laser off. Then turn on the laser and finish the alignment. Finally, turn the laser back off and move on to the next optic in the chain.

Since optomechanics are so varied, I shan't try to describe them all here. Browsing the catalogs of optical supply houses will give you an idea of the range of optics and optomechanics available. Suffice it to say that as a general principle, use the simplest component that will work. That's particularly true when it comes to motion control. If manual control will work, don't waste time installing and debugging a computerized controller. Also, design iteratively. Build the simplest possible version of your optics chain and use what you learn to improve it.

2.3 Measurement Error

For any quoted value to be useful, in any context, it must be accompanied by *some* kind of indication of its accuracy.[7] Even in casual conversation, it's typical for the approximate accuracy of a number to be implied by the context, by tone of voice, or by some other means. For scientific purposes, it's best to say explicitly what is the estimated error associated with a quantitative statement. Stating measurement errors explicitly also forces us to think about them explicitly, which leads to better science overall. So, the dictum to "always include error bars" annoying to many beginners used to a less formal approach, has a very good rationale. Just don't take it to unreasonable extremes. There's no need to record every possible source of error affecting a measurement but it is crucial that you identify and quantify *the dominant sources of error*.

There are two types of measurement error: Random errors, also called uncertainties, and systematic errors. Random errors are due to any uncontrolled factors that cause a measured value to change randomly between otherwise identical measurements. The histogram of measured values from a large number of identical measurements will generally be Gaussian. The standard deviation of the Gaussian is what is meant by the uncertainty. So, when we are estimating the uncertainty in a measurement, we are guessing at the standard deviation of the histogram that would result from a large number of measurements.

Random errors are introduced at many levels. In measurements taken by hand, where an analog scale is being read by a human, one source of random error is simply the process

---

[7] In this book, I use the words "precise" and "accurate" colloquially. There is a tradition in the English-speaking world of using the word "precise" to indicate the absence of random errors and the word "accurate" to indicate the absence of systematic errors. However, this usage is not universal and doesn't translate well to international readers. I try to avoid the usage and instead be explicit about the type of errors present, either random or systematic.

of reading the scale. When humans read a scale as carefully as they can, the scale-reading uncertainty is often about $1/5^{\text{th}}$ of the smallest scale gradation. Broadband electronic noise is a source of random error in most electronic measurement devices. For example, the exact time at which a photogate triggers as the light level approaches the predetermined trigger level, may be determined by random fluctuations of the light level causing the gate to trigger ever-so-slightly early or late. Even passive components like resistors have noise associated with the thermal motion of their constituent atoms and molecules, known as Johnson noise. Johnson noise is a form of a more general class of noise known as thermal noise, which afflicts everything from your high-end stereo to radio telescopes and gravitational wave detectors. It causes random fluctuations in these instruments' output that are not due to the signal being measured (a music track, a pulsar, binary black holes, etc.) Such fluctuations fall into the category of random error. Since random errors lead to histograms that are symmetric about the "true value," they are reported with in the $a \pm \Delta a$ notation. Unless stated otherwise, the Δa reported this way is assumed to represent only the random part of any error, and does not contain the systematic error.

Systematic error is due to some nonrandom effect on a measurement. A typical systematic error is an offset error. For example, if a small sliver has been cut off the zero end of a ruler, it will give erroneously high length measurements. Another typical systematic error, is a scale calibration error. For example, if a wooden ruler has expanded due to absorption of moisture, it leads to measurements that are erroneously low in proportion to the value being measured. As in the ruler example, offset errors and calibration errors can combine to make some parts of a measurement erroneously high and other parts erroneously low. Systematic error is not always constant. A laser with slow but periodic output power fluctuation may lead to unforeseen measurement errors that are systematic in nature and not easily recognized as such.

In general, we try to reduce both types of error during the design and design-iteration phases of an experiment. If a systematic error is identified during the operation of the experiment, it's normally expected that we either fix it or correct for it somehow. (For example, by recalibrating.) If systematic errors are known to be present and can't be fixed or corrected for some reason then they should be noted in the presentation of the final result. A sentence or two explaining the scope of the systematic error and why it could not be eliminated is appropriate in that case.

2.3.1 Propagating Uncertainties

When the measured quantities $a, b, \ldots$ are to be used in a calculation to obtain some derived quantity $f(a, b, \ldots)$ then uncertainty in the measured quantities will lead to a concomitant uncertainty in the derived quantity. If the measured values are

$$a = a_0 \pm \Delta a, \tag{2.2}$$

$$b = b_0 \pm \Delta b, \tag{2.3}$$

$$\vdots$$

Then Δf is obtained from Δa, Δb,... as follows

$$\Delta f^2 = \left(\frac{\partial f}{\partial a}\bigg|_{a_0,b_0} \Delta a\right)^2 + \left(\frac{\partial f}{\partial b}\bigg|_{a_0,b_0} \Delta b\right)^2 + \ldots \tag{2.4}$$

The partial derivatives are evaluated at the measured values of the parameters a_0, b_0,...
Equation (2.4) is only correct to first order. In other words, it is only useful for small Δa,
and Δb. I'll prove it for two measurement variables; the generalization for more variables
is straightforward.

Consider N measurements a_i of a and N measurements b_i of b. These are used to find
a set of N values f_i for f. If a_i and b_i are different from their true values by Δa_i and Δb_i,
respectively, then f_i will vary from its true value by Δf_i. To first order

$$\Delta f_i = \frac{\partial f}{\partial a}\bigg|_{\bar{a},\bar{b}} \Delta a_i + \frac{\partial f}{\partial b}\bigg|_{\bar{a},\bar{b}} \Delta b_i. \tag{2.5}$$

The partial derivatives are evaluated at the average values $\bar{a}$ and $\bar{b}$ of the N measurements of
a and b. To avoid clutter, we assume this to be the case below even though it's not indicated
explicitly. The mean-square deviation is

$$\begin{aligned}
\Delta f^2 &= \lim_{N\to\infty} \left[\frac{1}{N}\sum_{i=1}^{N} \Delta f_i^2\right] \\
&= \lim_{N\to\infty} \left[\frac{1}{N}\sum_{i=1}^{N} \left(\frac{\partial f}{\partial a}\Delta a_i + \frac{\partial f}{\partial b}\Delta b_i\right)\left(\frac{\partial f}{\partial a}\Delta a_i + \frac{\partial f}{\partial b}\Delta b_i\right)\right] \\
&= \lim_{N\to\infty} \left[\frac{1}{N}\sum_{i=1}^{N} \Delta a_i^2\left(\frac{\partial f}{\partial a}\right)^2 + \Delta b_i^2\left(\frac{\partial f}{\partial b}\right)^2 + 2\Delta a_i\Delta b_i\left(\frac{\partial f}{\partial a}\right)\left(\frac{\partial f}{\partial b}\right)\right].
\end{aligned} \tag{2.6}$$

Since Δa_i and Δb_i are *independent* and each randomly distributed about zero, the ratio of
the last term to the other terms will tend to zero for large N. (The cross term is negative
just as often as it's positive.) Also,

$$\lim_{N\to\infty} \frac{1}{N}\sum_{i=1}^{N} \Delta a_i^2\left(\frac{\partial f}{\partial a}\right)^2 = \left(\frac{\partial f}{\partial a}\right)^2 \lim_{N\to\infty} \frac{1}{N}\sum_{i=1}^{N} \Delta a_i^2$$

$$= \left(\frac{\partial f}{\partial a}\right)^2 \Delta a^2, \tag{2.7}$$

where Δa is the RMS[8] deviation of a (equal to the standard deviation for large N). Identical
relations hold for b. Substituting into Eq. (2.6) gives

$$\Delta f = \left[\Delta a^2\left(\frac{\partial f}{\partial a}\right)^2 + \Delta a^2\left(\frac{\partial f}{\partial b}\right)^2\right]^{\frac{1}{2}}. \tag{2.8}$$

---

[8] RMS stands for root-mean-square. The root-mean-square of a set of numbers $\{a_1, a_2, ..., a_N\}$ is
$\sigma = \sqrt{\frac{1}{N}\sum_{i=1}^{N} a_i^2}$.

As noted earlier, the partial derivatives are evaluated at the average $\bar{a}$ and $\bar{b}$ respectively. The best estimate of the average in a single measurement are the measured values a_0 and b_0. So, these partials should be evaluated at the measured values in the case of a single measurement. So, this equation is equivalent to Eq. (2.4).

Example 2.1 Error Propagation through a Product Propagating uncertainties in a and b to f when $f = ab$ using Eq. (2.4) gives

$$\Delta f^2 = (b\Delta a)^2 + (a\Delta b)^2$$
$$= a^2 b^2 \left(\frac{\Delta a^2}{a^2} + \frac{\Delta b^2}{b^2} \right).$$

So, for products

$$\frac{\Delta f^2}{f^2} = \frac{\Delta a^2}{a^2} + \frac{\Delta b^2}{b^2}. \tag{2.9}$$

Commonly used propagation formulas following from Eq. (2.4) are shown in Table 2.1. Note that most of the time only the dominant sources of uncertainty (the dominant terms in these formulas) need to be retained.

Table 2.1 Formulas for propagating uncertainties.

$f = a \pm b$	$\Rightarrow$	$\Delta f^2 = \Delta a^2 + \Delta b^2$
$f \propto ab$ $f \propto a/b$	$\Rightarrow$	$\frac{\Delta f^2}{f^2} = \frac{\Delta a^2}{a^2} + \frac{\Delta b^2}{b^2}$
$f \propto a^n b^m$	$\Rightarrow$	$\frac{\Delta f^2}{f^2} = n^2 \frac{\Delta a^2}{a^2} + m^2 \frac{\Delta b^2}{b^2}$

2.3.2 Testing Models against Data

In experimental physics, the fundamental question is:

"Does our understanding agree with observations?"

When our understanding is expressed as a mathematical relationship between observable quantities, we can test it in a rigorous way. We measure the physical quantities involved and check whether the expected relationships are evident.

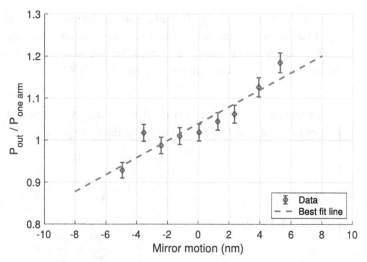

Figure 2.14 Simulated data for the relative power output from a Michelson interferometer versus mirror position. The dashed curve is a best-fit straight line.

An Example

Consider an experiment to check whether the electric field obeys the principle of linear superposition. To do this we could use a Michelson interferometer with monochromatic light of wavelength λ. Suppose x is the distance moved by an end mirror of the interferometer while y is proportional to the power seen at the output of the interferometer. We measure y as a function of x and compare the results to the expected response[9] assuming $x \ll \lambda$.

$$y = \left(\frac{4\pi}{\lambda}\right) x + 1. \tag{2.10}$$

The predicted relationship relies on the assumption that linear superposition of electric fields is exact. We now fit our measured data to a straight line: $y = a_1 + a_2 x$. If a straight line is not consistent with the data or if a_1 and a_2 are not consistent with the expected values, then something must be missing in our understanding.

Figure 2.14 shows a simulated data set for such an experiment with the best-fit line included. To evaluate whether the data is consistent with this best-fit straight line, we need to check how far the best-fit line lies from the data points and how that compares with the uncertainties. One expects the data to be scattered about the fit due to measurement uncertainties but a "chi-square test" (see below) can tell us whether the scatter is greater or smaller than expected. If the variation of the data around the fit is too large and we are sure we understand our apparatus and the errors, then we have to consider that the

[9] Example 5.1 presents an analysis of the Michelson response to mirror motion. All we need here is the expected response based on Eq. (5.9). In this example, I've chosen y to be the output power divided by the power at the output when the beam in one of the interferometer arms is blocked. That way the ratio can be measured directly on the single output photodetector. The output power when a single arm is blocked is half the input power to the interferometer, so $y \equiv I_{\text{det}}/I_{\text{one arm}} = I_{\text{det}}/I_{\text{in}}$. In terms of our variables, the response is twice the one shown in Eq. (5.9), x replaces ϵ and the sine is approximated by its argument.

fit function may be wrong. If that's the case, the physics that leads to the fit function may need to be examined as well. And that's ultimately what we want to discover when we do an experiment: whether or not the model representing our physical understanding agrees with the data. The lower the measurement uncertainties, the more stringently we can make that test.

I should emphasize here that *usually* it's not the physics that's found to be incorrect. Rather, there may be unaccounted for systematic errors that can explain a discrepancy with the model. However, if other, independent experimenters reproduce your result and also cannot identify any reason for the discrepancy, the case becomes stronger.

The steps required to do a fit, perform a chi-square test, and find the uncertainties in the fit parameters, are described below.

Finding the best fit

Let's assume we have taken N data pairs x_i and y_i, ($i = 1..N$) perhaps arranged with uncertainties Δy_i into a table as follows.[10]

x	y	Δy
x_1	y_1	Δy_1
x_2	y_2	Δy_2
⋮	⋮	⋮
x_N	y_N	Δy_N

Our task is then to evaluate whether the relationship between the x_i's and the y_i's is actually well described by our chosen fit function $f(x, a_1, a_2, ..., a_m)$. The idea is to vary the parameters $a_1, a_2, \ldots, a_m$ so that the mean square vertical distance between the data points and the fit function becomes as small as possible. For this reason, we call it a "least squares" fit. However, we'd like data points with low uncertainty to count more in defining the best fit than data points with large uncertainty. So, we will divide the vertical distance of each data point from the fit curve by the uncertainty Δy_i thus reexpressing the distance in units of the uncertainty

$$d_i \equiv \frac{y_i - f(x_i, a_1, a_2, ..., a_m)}{\Delta y_i}. \tag{2.11}$$

The best fit is then given by minimizing the sum of the squared distances

$$\chi^2 = \sum_{i=1}^{N} d_i^2. \tag{2.12}$$

This quantity is known as the chi-square, χ^2, and is pronounced "Kai-square."

---

[10] For this introduction, we assume that the uncertainties in y dominate and we can ignore the errors in x. (If x−errors dominate, we can always exchange the axes.) The case where the uncertainties in x and y are similar in magnitude requires either an extension of the method or a conversion of x−errors into y−errors using the slope of a preliminary fit.

A related function is known as the "reduced chi-square" and it's just the chi-square divided by $N - m$ where N is the number of data points and m is the number of fit parameters.

$$\chi_{\text{red}}^2 = \frac{1}{N - m} \sum_{i=1}^{N} d_i^2.$$ (2.13)

It doesn't matter whether you choose to minimize the chi-square or the reduced chi-square since they are only different by a constant. Note that for a large number of data points, the reduced chi-square is just the mean square distance between the data and the fit, in units of uncertainty. If deviations of the data points from the fit are only due to uncertainties, then the reduced chi-square should be very close to 1. So, we evaluate a fit by asking

$$\boxed{\text{"Is } \chi_{\text{red}}^2 \approx 1 ?\text{"}}$$

The denominator $N - m$ in Eq. (2.13) allows us to compare fits between model functions with *different numbers of fit parameters*. You may recall that given any N points, we can always find an Nth order polynomial that passes through all the points. So, an Nth order polynomial will always provide a "perfect fit" to set of N points. But such a fit does not represent the underlying physics and should not be considered "good." After all, it's just chasing random noise. An Nth order polynomial fit has $m = N$ coefficients. So, $N - m = 0$ and $\chi_{\text{red}}^2 \rightarrow \infty$. According to the above-mentioned chi-square criterion, it is indeed a terrible fit. Other things being equal, the presence of $N - m$ in the denominator biases us to prefer fit functions with fewer parameters – a rather nice implementation of Occam's razor!

Discussion

The reduced chi-square is a function of only the parameters a_i since the data points have been summed over. Writing it out again, explicitly

$$\chi_{\text{red}}^2 (a_1, a_2, \ldots, a_m) = \frac{1}{N - m} \sum_{i=1}^{N} \frac{[y_i - f(x_i, a_1, a_2, \ldots, a_m)]^2}{\Delta y_i^2}$$ (2.14)

It can be convenient to think of the reduced chi-square as a surface over the $a_1, a_2, \ldots, a_m$ axes. Fitting is then the process of finding the location $(a_1', a_2', \ldots, a_m')$ of the minimum of the surface. (The visualization of χ_{red}^2 as an actual surface over a plane only works literally for two-parameter fits. In the case of fits to more than two parameters, χ_{red}^2 is a hypersurface.)

Efficient fitting routines have been developed that basically follow the χ_{red}^2 surface downhill until a minimum is reached. Of course this risks only finding a local minimum rather than the global minimum, so good starting guesses for the parameters should be provided to the fit routine. In the special case where the fit function $f(x_i, a_1, a_2, \ldots, a_m)$ is linear *in the parameters*, there will be only one minimum of the χ_{red}^2 and its location can be calculated analytically. I won't describe the details of either of these methods here. While it's important to understand *what* a curve-fitting routine is doing and why, writing one's own is not really necessary. Excellent curve-fitting routines are available for Matlab, Python, and other languages. Appendix B.3 contains example code in Matlab for obtaining a fit similar to the one shown in Figure 2.14.

So, does one ever need to actually plot the χ^2_{red} surface? Most likely, you will find yourself plotting it before very long (or plotting 2D *cuts* through the minimum if it's a hypersurface). The most common reason for plotting the χ^2_{red} is to evaluate any interdependence between the best-fit values of parameter pairs. Interdependence of two parameters will show up as a diagonal valley in the χ^2 surface, indicating that some linear combination of the parameters is poorly constrained by the fit.

As discussed earlier, when the fit reflects the underlying physics well, variations of the data around the fit are purely due to measurement uncertainties and so $\chi^2_{\text{red}} \approx 1$. The spread of data points around the fit is random and the standard deviation of those variations equals the uncertainty. In other words, about two thirds (68%) of the error bars should be expected to touch the best fit line. So, it's possible to glance at data with error bars and make a quick judgement of the applicability of the model by estimating whether fewer than two thirds of the error bars touch the fit.

Chi-square test

We know that a value of χ^2_{red} significantly greater than one implies that the fit-function is likely to be inconsistent with the data. However, we'd prefer to make a quantitative statement about the *probability* that the fit-function is consistent with the data based on the actual value of χ^2_{red}. Under the assumption that the errors are Gaussian, and assuming our fit function is correct, it's possible to construct a function $P_\chi(X, v)$ that will tell us the probability of getting a value of χ^2_{red} greater than or equal to X due to random-error fluctuations alone. $P_\chi(X, v)$ takes both the value X of the reduced chi-square and the number of degrees of freedom $v = N - m$. Finding the probability $P_\chi(X, v)$ based on the χ^2_{red} and v for a particular fit to 102 data points is known as applying the chi-square test. For example, if we perform a two-parameter fit and find that the value of the reduced chi-squared is 1.243, then $P_\chi(1.243, 100) = 0.05$. The result of this chi-square test is that there is only a 5% chance that the fit-function is consistent with the data.

The calculation of $P_\chi(X, v)$ is described in statistics books and the values are tabulated. For example, you can use Table C.4 in Bevington (2003). (Online calculators also exist but make sure you find the right one.) To use a chi-square test table, you'll need the number of degrees of freedom, $v = N - m$ and the value of χ^2_{red}. The table will then give $P_\chi(X, v)$.[11]

So far we've mostly concentrated on χ^2_{red} values above one because these indicate that the fit is too far from the data points. If your χ^2_{red} is significantly less than one, it generally indicates that you've over-estimated your errors. Your fit is "too good" given the uncertainties you're assigning. This doesn't serve to rule out your model but indicates that your data is not varying as much as one would assume given the assigned uncertainties.

---

[11] For example, using Table C.4 in Bevington (2003): With straight line fit to 102 data points, $v = N - m = 100$. Assuming we got a $\chi^2_{\text{red}} => 1.243$, then our fit function has only a 5% chance of being correct.

As you can see from any chi-square test table, the probability of any particular χ^2_{red} value depends strongly on the number of data points you have taken. If you have many data points, say hundreds or thousands, then χ^2_{red} should be very close to 1. Even minor increases from one indicate a problem with the fit. Other things being equal then, more data will give you a more stringent chi-square test of your model. Also, since we're dividing by the measurement uncertainties to get the χ^2_{red}, smaller measurement uncertainties mean that less deviation from the fit function is allowed. As one might expect, lower measurement uncertainties and a larger number of independent data points leads to a more stringent test of the underlying physics.

Finding the Uncertainty in the Fit Parameters

Since the underlying data (x_i, y_i) have uncertainties, the best-fit values $(a'_1, a'_2, \dots a'_m)$ of the parameters will have corresponding uncertainties $(\Delta a'_i, \Delta a'_2, \dots a'_m)$. To estimate the uncertainty in the fit parameters, we need the chi-square, Eq. (2.12), rather than the reduced chi-square. A plot of χ^2 versus any one of the parameters a_i while fixing the rest at their best-fit values should always show a minimum at the best-fit point, a'_i. This is the overall minimum of the chi-square surface and the minimum value is $\chi^2_{min} = \chi^2(a'_1, a'_2 \dots a'_m)$ at this point. *The uncertainty in a_i is the amount by which you must vary a_i from its best fit value in order for the χ^2 to increase by one.* Figure 2.15 shows an example of this. The upper panel shows the chi-square surface corresponding to a two-parameter, straight-line fit similar to the fit shown in Figure 2.14. The dashed white contour in the upper panel indicates the locations (a_1, a_2) at which $\chi^2(a_1, a_2) = \chi^2_{min} + 1$. In other words, the contour shows the distance from the minimum at which the chi-square has risen by 1. The horizontal distance from the minimum to the contour is the uncertainty $\Delta a'_1$ in the best fit value a'_1. Similarly, $\Delta a'_2$ is the vertical distance from the minimum to the contour. The lower panel shows the cut along the long-dashed, horizontal line that corresponds to allowing a_1 to vary while keeping a_2 at its best-fit value, a'_2. The cut shows the minimum chi-square at the best fit values of the parameters and the corresponding uncertainty $\Delta a'_1$ is easy to read directly from the graph.

As demonstrated by Figure 2.15b, it's not necessary to calculate the full chi-square surface to find the uncertainties. One only needs to calculate the chi-square cuts. Generating these cuts is a fairly easy way of obtaining the uncertainty in the parameters. It's a good way to get a handle on fitting uncertainties, especially to begin with. However, if you do a lot of fitting, generating chi-square cuts soon gets old. Since the cuts are through a minimum of the chi-square surface, they can almost always be approximated by tangent parabolas. The uncertainties in the fit parameters are then estimated by calculating where these tangent parabolas have increased by 1. The tangent parabolas can be found from the curvature of the chi-square cuts at the minimum. Some fit routines will use this information to estimate the uncertainties in the parameters for you. Other fit routines may return the curvature of the cuts at the minimum or some related quantity from which you can find the parameter uncertainties. The code in Appendix B.3 shows an example how to make a direct estimate of the parameter uncertainties based on information about the chi-square curvature.

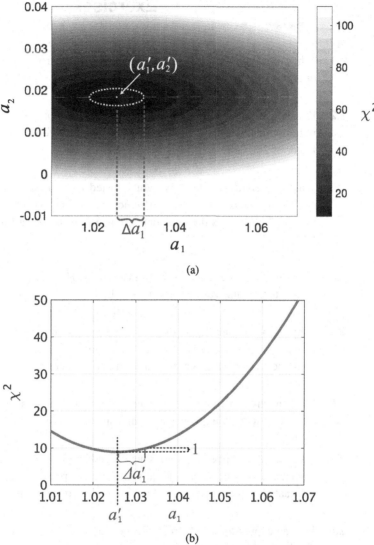

Figure 2.15 **(a)** The χ^2 surface around the best-fit. The best fit values a_1' and a_2' are indicated by the small white dot in the darkest (lowest) region of the surface. The 1-sigma uncertainty is indicated by the white contour. The uncertainty $\Delta a_1'$ is the distance from the best fit to the contour along the a_1-axis. **(b)** A cross section through the surface shown in (a). The cross section is taken by cutting the surface by a plane perpendicular to the page, parallel to the a_1–axis and passing through the χ^2 minimum. The cut is indicated by the faint wide-dashed line in (a). As shown, the uncertainty in a_1' can be found directly from this cut more easily than from the full surface plot.

Exercises

2.1   When *must* you wear laser safety glasses? Answer from memory. If you get it wrong or aren't sure, reread Section 2.1.

2.2   If laser glasses are only available in integer-valued optical densities, what is the minimum optical density of laser glasses required to reduce the transmitted power by a factor of 200,000?

2.3   There are eight "good safety habits" listed in Section 2.1. Recite as many as you can from memory. If you can't remember at least six, reread the section and try again. Also, what is the most dangerous non-beam hazard in your particular optics laboratory?

2.4   Find an actual measured spectrum of any QTH lamp (online is fine). Estimate the lamp's temperature and justify the result.

2.5   Why is the filament in Figure 2.1 bright in the middle but dim at the ends? The proximate cause is fairly obvious but try to explain how the proximate cause itself could occur if the filament is of uniform cross section.

2.6   Explain why a thermopile can accurately measure the total power in a beam of broadband light with an unknown spectral distribution but a photodiode cannot.

2.7   Calculate the number of photons per second leaving the aperture of a 1mW laser with wavelength $\lambda = 632.8$ nm. What is the average power, in Watts, of this beam if it is attenuated so that an average of only one photon leaves the aperture per second?

2.8   Find an expression for the relative uncertainty $\frac{\Delta f}{f}$ in the focal length f in terms of a, Δa, b, and Δb, when f is given by

$$\frac{1}{f} = \frac{1}{a} + \frac{1}{b}. \tag{2.15}$$

2.9   Show that if $f = N^{ka}$, where N and k are constants, then $\frac{\Delta f}{f} = |k \ln N| \Delta a$.

2.10 Show that if f is related to a by the formula for decibels: $f = 10 \log_{10}\left(\frac{a}{a_{\text{ref}}}\right)$ where a_{ref} is a constant, then $\frac{\Delta f}{|f|} = \left|\ln\left(\frac{a}{a_{\text{ref}}}\right)\right|^{-1} \frac{\Delta a}{|a|}$.

2.11 If the fractional uncertainty in x is 3% then what is the fractional uncertainty in x^n? Is your result correct for both positive and negative values of n?

2.12 Fit the data below to a straight line: $y = ax + b$. What are the best-fit values of the slope a and the intercept b? Is the fit function consistent with the data? By plotting the (unreduced) chi-square cut through the minimum parallel to the a–axis, find the uncertainty in the slope. (*Hint*: Leave b at the best-fit value and calculate the chi-square as a function of a alone.)

x	0.8	2.0	3.0	3.8	5.1	6.1
y	4.3	5.6	12.3	13.0	18.7	19.1
Δy	1.5	2.1	1.3	2.3	2.8	1.6

2.4 Experiment: Components of an Optics Chain

Objectives

1 Set up a well-aligned optics chain.
2 Use an amplified photodiode and CCD/CMOS camera.
3 Measure the reflection characteristics of different mirrors.
4 Investigate clipping, vignetting, and optical throughput with a lens relay.
5 Perform nonlinear least-squares fits.
6 Gain experience with reflective prisms.

Equipment Needed

- Two identical, ~25 mm diameter, first-surface, flat, aluminum mirrors.
- One first-surface, flat, gold mirror.
- One laser-line dielectric mirror in the visible range.
- A bright broadband light source, such as a halogen lamp and a set of about ten optical band-pass filters throughout the visible and near-infrared. Alternatively, a set of about ten narrow-band LEDs with a selection of wavelengths throughout the visible and near infrared. If a monochromator is available, it can be used instead.
- A variable iris-style aperture.
- A bright, rear-illuminated frosted screen to serve as the "object." If that is not available, a frosted light bulb will work almost as well.
- Black tape or foil with which to mask the object.
- Two identical lenses with focal lengths, $f \sim 50$ mm and lens diameter about 25 mm.
- Two identical lenses with focal lengths, $f \sim 50$ mm and lens diameter about 50 mm.
- One vertical knife edge on a mount that allows it to be slid into or out of the beam easily.
- A photodiode, photodiode amplifier, and multimeter to read it out.
- Laboratory grade CCD/CMOS camera.
- A variety of reflective prisms. For example: pentaprism, amici roof prism, dove prism, trihedral prism (retroreflector).

Reflection from First-Surface Mirrors

Flat, first-surface mirrors are a crucial component of optics chains. Choosing the correct mirror type can have a big effect on the throughput and other properties of your optics chain. Your first task is to measure the reflectance R, reflected power divided by incident power, at close to normal incidence of: a gold-coated mirror, an aluminum mirror, and a laser-line dielectric mirror. Plot and compare the measured curves $R(\lambda)$ for the three mirrors. Next, choose a single wavelength where the laser-line mirror has good reflectivity. Measure it's reflectance as a function of angle. Do the same for one of the metallic mirrors and compare the results.

> *Recommendations*: Make sure your photodiode amplifier is not close to saturating during the measurement. You can usually find out what the saturation point is by exposing the sensor to a bright light source. Also, the output voltage typically saturates when it

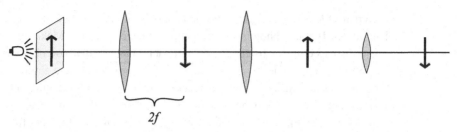

Figure 2.16 Top view of a lens relay. The leftmost arrow, within a window, represents the illuminated object (which in the actual experiment setup is simply a small rectangular aperture with no arrow). For clarity, the object's size has been exaggerated. All image-lens distances are $2f$ as shown.

gets within a volt or two of the DC voltage being used to power the photodiode amplifier. Make sure you measure the optical power immediately before and after the coating. The incident beam should be collimated to the extent possible. You can collimate the input beam by placing the light source at the focal point of a lens. To avoid clipping, the collimated beam should be significantly smaller than the mirror whose reflectivity you are measuring. You can accomplish that by using an aperture after the collimating lens. You should also use a short focal length, large diameter lens to focus the light onto the active area of the sensor. That way, you can be sure there is no clipping at the edges of the sensor. Note that while it is also possible to do this measurement using a spectrometer, part of the goal is to give you experience using a discrete psuedo-monochromatic light source and a photodiode.

For the transmission versus angle of incidence, remember that the angle of incidence is half of the total angle through which the beam is reflected. There may be rotating platforms available for your use that are graduated in degrees. If not, use a ruler and trigonometry to get your angles.

Lens Relay

The purpose of this section is to introduce you to setting up a short optics chain and to investigate the properties of a lens relay system. This sort of system can be used whenever an image needs to be transported a short distance. Such systems are used in periscopes, in astronomy to bring an image from a telescope to an instrument, in photography, in endoscopes, and so forth. Here you will set up a very simple system where all the lenses have the same (or similar) focal length.

Set up the lens relay shown in Figure 2.16 using two 50 mm lenses and one 25 mm lens. Use the black tape to mask the frosted screen so that only a "square" window is illuminated, $\sim 7 \times 7$ mm. This is your object. The object should not be larger than the sensor chip in your CCD/CMOS camera; if necessary it can be made smaller. Mark one of the corners of the object so that you can tell its orientation. The purpose of the lens relay is to create an image of this object at locations spaced every $4f$ along the beam path. As you set up the relay, use a piece of card as a screen to check that an image is formed halfway between each pair of lenses and after the last lens. Using the bare CCD/CMOS camera (no camera lens attached) obtain photographs at the three image locations. Make sure the actual image sensor is placed at exactly the location of each image so that you obtain photographs that are in the best possible focus. It's important to keep the camera settings the same for all three images and make sure the exposure level is not being set automatically. Import the images into a computer. Choose an identical cut through each image (see, for example, Appendix B) and plot the irradiance along the cut. Display the cuts from all the images on the same graph to show how the image irradiance changes as a function of position in the chain. Explain the rate of irradiance drop between images. If you see evidence of vignetting, speculate on the reason for it. Now exchange the middle (50 mm) lens with the last (25 mm) lens. How does that change the irradiance of the images (if at all)? Find a plausible reason for any changes or lack thereof.

Now remove the last two lenses and move the first lens somewhat closer to the source and view the image it produces on a card. The image will grow and move farther from the lens as you move the lens closer to the source. Set the lens position so that the image diagonal is roughly 20–25 mm in length. Now, use the 50 mm diameter lens to relay the image formed by the first by a distance $4f$ as before. Note the size and shape of the downstream image. Capture a photograph. Then replace the 50 mm diameter lens with a 25 mm diameter lens. Capture a photograph. How does the image compare with the one produced by the 50 mm lens. Explain the reasons for any changes or lack thereof.

Recommendations: Before beginning, decide on the beam height and make sure that all the components you intend to use can be placed at the appropriate height. Place bases and post-holders in their rough positions first. Then work down the chain starting at the light source and place the posts with the lenses already attached into the post-holders. Keep the optics chain as straight as possible, ensuring that the beam remains at the same height and follows a straight line on the optical table. (You can use the hole-patterns for reference.) Make sure you familiarize yourself with the camera settings before taking pictures. There should be a way of fixing the light settings (exposure time, gain, etc.) so that images can be quantitatively compared. In other words, the pixel values must be proportional to the irradiance by the same factor in all the images.

Partial images

In this section, we will use a knife edge to clip the beam and observe how the result depends on the location of the knife edge relative to the image location. Set up the system as shown in Figure 2.17a. Place the knife edge at the location of the intermediate image as shown and slowly move the knife edge into the image until it's about halfway into the beam. Capture a photograph. Repeat the process using the same camera settings but move the knife edge so that it immediately follows the first lens, as shown in Figure 2.17b. Repeat for several intermediate locations of the knife edge between the lens and the image. In each case, also make a note of how the qualitative features of the second image depend on how far the edge is inserted into the beam.

Finally, choose a location for the knife edge that is near the image but sufficiently upstream from it so that the knife edge is in poor focus on the camera. (The edge of the knife edge should be a blur that covers about a quarter of the image.) Take a photograph with

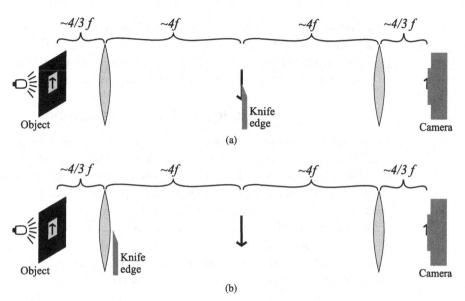

Figure 2.17 Blocking part of the image. **(a)** Knife edge placed about halfway into the beam at the first image location. **(b)** Knife edge placed about halfway into the beam immediately after the first lens.

the knife edge in this position and extract a cut from the image that lies perpendicular to the out-of-focus knife edge. Take an identical reference photograph with the knife edge entirely removed. Extract an identical cut from the reference photo and use it to normalize the cut from the first photograph. Plot the normalized irradiance versus position along the cut. Now discard parts of the cut that do not contain any shadow from the knife edge. The portion of the cut having some shadow from the knife edge will constitute your data. Develop a simple but physically motivated model for the shape of the irradiance versus position in the shadow. Fit your model to your data. Without doing a full chi-square analysis, decide whether the fit to your data is acceptable. Make a note of your reasoning.

> *Recommendations*: You'll need to think about the nature of errors in the camera images. It may be a good idea to repeat this part several times while intentionally varying things like the precise alignment of the optics chain, camera placement, and so forth. That will give you a feel for the actual variation between images in nominally identical measurements.

Image Parity

Reflective prisms are sometimes used in lieu of mirrors to achieve a reflection with a particular parity along one or more axes. The parity for a particular axis is negative if the image is inverted along that axis; otherwise it's positive. Investigate the prisms you have by placing them between your first lens and a screen. Record the effect on the parity of the image along each transverse axis. Do the same for a simple mirror. If you have a dove prism, also try rotating it about the optic axis.

Recommendations: To evaluate the parity of an image along a particular axis, view it from the same direction as the object is viewed. So, if the object is viewed from downstream, the image should also be viewed from downstream (which could be done in practice if the image was formed on a ground-glass screen).

Ideas for Further Investigation

If a thermal detector with a flat response is available, measure the response of a silicon photodiode by comparing it with the thermal detector. At each available wavelength, take the ratio of the silicon photodiode response to the thermal detector response. If an InGaAs (indium-gallium-arsenide) or Ge (germanium) photodiode is available, also measure its spectral response at the wavelengths you have available.

Compare the theoretical response of a laser mirror consisting of a quarter-wave stack of dielectric coating layers, Eq. (5.14), to the spectral response you measured. Do the theoretical and measured spectra agree qualitatively? Is the theoretical reflectivity in the high-reflectivity band consistent with what you measured?

3   Geometric Optics

3.1 The Geometric Optics Approximation

In geometric optics, we assume that the wavelength of light approaches zero. In that limit, successive wavefront normals trace out straight lines in free space, just like a stream of classical particles. For example, laser beams (Gaussian beams) spread out slowly as they travel, due to diffraction. In the limit $\lambda \to 0$, this spread also becomes zero and all the light moves in a perfectly straight line. (See, for example, Eq. (4.15).) We imagine these straight line paths as "rays" of light that are affected only by the optics placed in their path.

In geometric optics, the light waves are also assumed to be incoherent so they will have random phase and amplitude. The mean square field amplitude resulting from a sum of such random fields is the sum of the mean squares of the contributing fields. $E_{tot}^2 = E_1^2 + E_2^2 + \ldots$ Just like random errors, random fields add in quadrature. Irradiance (power per unit area) is proportional to the field squared, so this implies that the total *irradiance* at any point in space is just the sum of the contributing *irradiances*: $I_{tot} = I_1 + I_2 + \ldots$ That's of course what our intuition tells us should happen. Since we are used to incoherent light in everyday life our intuition is attuned to that case. It is the coherent case that's less intuitive.

When diffraction is insignificant and coherence is low, geometric optics provides a good model for the behavior of light. The most important application of geometric optics is in imaging. Yet, geometric optics is applied in many other contexts such as architectural lighting design, microwave antenna design, spectroscopy, and so forth.

3.2 Refraction

Refraction refers to the tendency of light to change direction when there is a change in the index of refraction of the medium in which the wave travels. The index of refraction n of a medium is

$$n = \frac{c}{v}, \tag{3.1}$$

where c is the speed of light in vacuum and v is the speed of light in the medium. Since $v \leq c$, the index is always greater than or equal to one. For linear media, refraction is governed by Snell's law. Consider a ray of light passing between two regions of different indexes of refraction. The interface is planar and the angle between the interface's surface

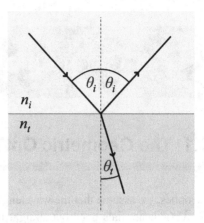

Figure 3.1　Construction for Snell's law.

normal and the incident ray is known as the angle of incidence (see Figure 3.1). The angle of transmission θ_t is related to the angle of the incidence θ_i by Snell's law

$$n_t \sin \theta_t = n_i \sin \theta_i. \tag{3.2}$$

Example 3.1 Thin Lenses　Consider the plano-convex lens shown here. It is supposed to represent a *thin* lens but it's thickness and curvature are greatly exaggerated to illustrate the angles involved.

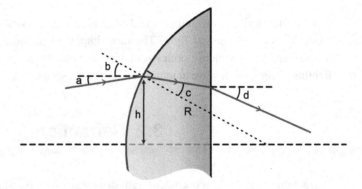

In the thin lens approximation, the orientation of the lens does not affect the behavior. However, in order to reduce aberrations in the real world, it's usually advantageous to place the convex side on whichever side the beam is more collimated. For example, when focusing a collimated beam, the curved side should face upstream as shown here. A ray incident on the curved surface of the lens makes an angle a with the horizontal and $a + b$ with the surface normal. After refracting, the angle of the ray (now inside the lens) is c with respect to the surface normal. After refracting at the planar surface, the ray leaves the lens at an angle d with respect to the horizontal. The radius of curvature of the lens is R, and the angle between the radius of curvature and the horizontal is b at the point where the ray first enters the lens. Provided $a \ll 1$ and R is large compared to the diameter of the lens (the lens is "thin") then the sines in Snell's law can be approximated by their arguments. If the lens with index n is in air, or in vacuum, ($n_i \approx 1$), Snell's law applied to the two interfaces gives

$$cn = (a + b) \tag{3.3}$$

$$\text{and} \quad d = (b - c)n. \tag{3.4}$$

Since the lens is thin, its radius of curvature is large compared to its diameter, so $b \approx \frac{h}{R}$ where h is the height at which the ray enters the lens. The angle $\Delta\theta$ through which the lens turns the ray is

$$\Delta\theta = a + d = \frac{n-1}{R}h. \tag{3.5}$$

So, *the action of a thin lens is to turn a ray through an angle proportional to the height at which the ray hits the lens.* By definition, a ray entering parallel to the optic axis, should pass through the focus a distance f from the lens. So, $\Delta\theta \approx \frac{h}{f}$. Comparing this with Eq. (3.5) reveals the relationship between the radius of curvature of a plano-convex lens and its focal length.

$$f = \frac{R}{n-1} \tag{3.6}$$

$$\approx 2R \quad \text{(for glass lenses in air).} \tag{3.7}$$

Since Snell's law applies to any interface, another way of stating Snell's law is to note that the quantity $n \sin\theta$ is unchanged across any planar interface.

$$n \sin\theta = \text{Constant.} \tag{3.8}$$

This holds for any interface including ones where the change of n across the interface is infinitesimally small. In other words, this holds for continuous changes in n where θ is measured with respect to $\vec{\nabla}n$. Differentiating Eq. (3.8) with respect to n leads to

$$\frac{d\theta}{dn} = -\frac{1}{n}\tan\theta. \tag{3.9}$$

If we choose our axes so that $\hat{x} \propto -\vec{\nabla}n$ and substitute $y' = \frac{dy}{dx} = \tan\theta$, we obtain a differential equation determining the path of the ray

$$y'' = -y'\left(1 + y'^2\right)\frac{n'}{n}, \tag{3.10}$$

where primes refer to differentiation with respect to x. Figure 3.2 shows an extreme example of the bent path taken by a ray when the index falls off quickly with height. For the purposes of the figure, we chose a relative change in index per unit height to be $n'/n = 0.01$ (distance units)$^{-1}$. Note that the ray turns toward higher index. Since the index of refraction of air depends on its density, light rays passing through the atmosphere exhibit similar but less severe bending. The apparent location of stars are therefore not

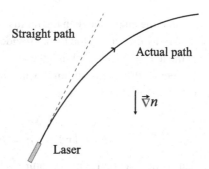

Figure 3.2 Actual path of a ray as compared to a straight line path when the index of refraction
falls off with height. The path satisfies Eq. (3.10). (In particular this case, the x-axis
points straight up and the positive y-axis points to the right.) The index has an
inverse-exponential dependence with height. We chose a constant relative rate of
index decay of -1% per unit distance and the height reached by the ray shown here is
about 80 units of distance.

exactly where they would be if viewed from orbit. Typical deviations are in the range of
minutes of arc. A more dramatic exhibition of this effect occurs above hot asphalt when the
shimmering, water-like image of the bright sky is seen refracting from the road ahead. This
is sometimes called a "mirage," which is derived from the same root as the word "mirror."
In that case the rays travel along an upwardly concave path because the index is lower near
the hot road than in the cooler air higher up.

3.3 Basic Imaging

Every point on a typical illuminated object scatters light in all directions. For the purposes
of imaging then, an illuminated object can be considered to be a source of incoherent
rays traveling out in all directions from every point on the object, unless blocked. As an
idealization, we consider the object to lie in a plane – the source plane. If all rays emanating
from any given point in the source plane are coincident on a corresponding single point in
some downstream field plane, we have an imaging system. The result is that the *relative*
orientation of points in the source plane is reproduced in the field plane with only a change
in overall size and orientation. The simplest imaging system is a single lens or curved
mirror (see Figure 3.3). The shape of the lens or mirror surface should ideally be parabolic.
Parabolic optics, also known as "aspherics," are considerably more difficult and expensive
to manufacture than optics with spherical surfaces, so we usually make do with spherical
lenses and mirrors. As long as the radius of curvature of a spherical optic is significantly
larger than its diameter, the spherical surface figure won't deviate much from that of a
parabola with the same radius of curvature at the center. The use of spherical optics does
induce errors, known as spherical aberrations, but they are usually subtle and hard to see
in the image produced. When the lens used for image formation is "thin" (thickness much

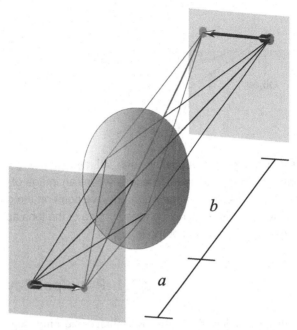

Figure 3.3 An *object* in the source plane – foreground arrow – is imaged by a single lens onto the field plane forming an *image* that is inverted and magnified. Representative rays are shown emanating from two different points on the object. Note that all rays from each point on the object converge at a single point in the field plane. The rays are only convergent in the single field plane shown. If we were to draw another field plane farther from, or closer to, the lens, the rays would not be convergent in that plane. The image in all such planes would be "out of focus." The image is in focus at a particular distance from the lens and nowhere else.

smaller than it's diameter) and all rays make fairly small angles with the optic axis, the location of the image and object with respect to the lens is given by the so-called thin lens equation

$$\frac{1}{f} = \frac{1}{a} + \frac{1}{b}. \tag{3.11}$$

The distance from the object to the lens is a, the distance from the lens to the image is b and the focal length of the lens is f. Note that for objects at infinity, the image forms at $b = f$ but for objects closer than infinity, the image forms at $b > f$. If the object is inside the lens' focal length, $a < f$, no positive solution for b exists and no image can be formed. If an image is formed, the ratio of the size of the image to the object is known as the magnification, M. Referring back to Figure 3.3 and considering the rays passing through the center of the lens, similar triangles on each side of the lens imply that the relative lengths of the arrows are

$$M = a/b. \tag{3.12}$$

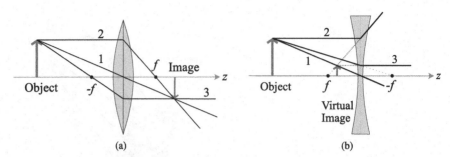

Figure 3.4 **(a)** Principal rays for a positive lens. All rays emerging from a given point on the object converge on the far side of the lens where an image of the object is formed. (Here the arrow tip was chosen as a representative point on the object.) **(b)** Principal rays for a negative lens. Now all rays on the far side of the lens appear to emanate from a virtual image on the near side of the lens.

3.3.1 Principal Rays

While a single lens does provide an image, most imaging systems are comprised of multiple lenses. A nice graphical way of tracing the progression of an image through a system of lenses is to use so-called principal rays. They are simply three rays whose path is easy to plot. For example, the ray that passes through the center of a lens doesn't change its path due to the lens, so it is one of the principal rays.

Figure 3.4a shows the three principal rays for a positive lens. The first principal ray passes through the center of the lens. The second principal ray is parallel to the optic axis on the near side of the lens and passes through the focal point of the lens, f, on the far side of the lens. The third principal ray is parallel to the optic axis on the far side of the lens and must therefore pass through $-f$ on the near side of the lens. Note that the plane in which the principal rays converge is the image plane, where all rays from a given point on the object converge.

Figure 3.4b shows the equivalent principal rays for a negative lens. In this case, the order of the foci is switched. Since f is a negative number, $-f$ is now on the downstream (positive) side of the lens. In the case of a negative lens, no actual image is formed. However, the rays emerging from the lens appear to be emerging from an image on the same side of the lens as the object. Such an image is called a virtual image. In the figure we just see the principal rays appearing to emerge radially from the tip of the arrow's virtual image. There are however, an infinite number of such rays, all of which appear to emerge radially from the arrow's tip. The same holds for any other point of the virtual image. In the case of the negative lens, the first principal ray is unchanged from the case of the positive lens. The second principal ray is parallel to the optic axis on the near side of the lens, as before. On the far side of the lens it turns away from the optic axis but extending it backward through the lens causes it to pass through f (as it does in the positive case except f is now on the near side of the lens). The third principal ray is parallel to the optic axis on the far side of the lens, as before. The ray on the near side of the lens can be extended to pass through $-f$ and this sets the height of the emerging ray.

It's really only necessary to use two of the three principal rays. Which two you choose depends on the situation but most often it will be rays 1 and 2.

3.3.2 A Simple Model of the Eye

We can use principal rays to illustrate the basic function of the most important optical instrument – the human eye. The eye is a very complicated biological structure whose function is fundamentally that of a movie theater for the brain: The lens and cornea at the front of the eye project an image of the illuminated world onto a curved screen at the back of the eye, called the retina. There, the image is recorded by photosensitive cells (rods and cones) and relayed to the brain. The highest resolution vision is at the center of the retina, in the region known as the fovea. There, photoreceptor cells (mostly cones) are spaced only a few microns apart. The cornea is the transparent front part of the eye (onto which contact lenses are placed). The cornea sits in front of the pupil, which is the opening in the iris. Immediately inside of the pupil sits the lens of the eye. The cornea provides roughly two thirds of the refractive power of the eye while the lens provides the remaining third. The lens is used to vary the focal length of the eye using the ciliary muscle, an annular muscle surrounding the lens. When the ciliary muscle is relaxed, the lens is stretched flat and thin and has a long focal length appropriate for imaging objects at infinity onto the retina. When the ciliary muscle is tightened, the lens thickens and curves, giving it a shorter focal length. We will model the lens and cornea combination as a single effective lens located at the rear surface of the actual lens and having variable focal length between about 14 and 17 millimeters.[1] As a result, the eye can form an image on the retina for an object at any distance between about 7 cm with the ciliary muscle fully tightened and infinity with the muscle relaxed.[2]

Figure 3.5 shows a simple schematic eye imaging each of two similar objects at different distances from the eye. (Only one of these objects is in focus at a given time.) As shown in the figure, the orientation of the image on the retina is opposite that of the object being viewed. In each case, two principal rays converge at the retina, showing that an image is indeed formed on the retina and the focal length of the eye *must change* to facilitate this. The focal length required for a near object f_2 is shorter than that required for a distant object f_1. (The figure is somewhat exaggerated in the sense that the object-eye distances are too small for a normal human eye to accommodate but chosen so that the retinal images are of a reasonable size for illustration.) We also see that similarly sized objects at different distances form images of different sizes on the retina. Thus, an object appears smaller as it is moved farther away. At the closest distance the eye can accommodate, it can resolve details as small as a few tens of microns.

3.3.3 Magnifying Devices

The magnifying glass, microscope, and telescope are all examples of magnifying devices that use lenses (or equivalent mirrors) to achieve their function. Each of these devices is

---

[1] See, for example, Pedrotti (2007) or Hecht (2017), which also give much more detail.

[2] The minimum distance of 7 cm corresponds to a well-functioning eye in a young person. The minimum distance that the eye can accommodate increases with age.

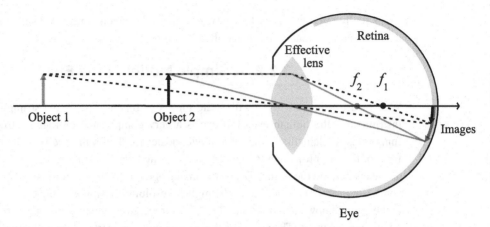

Figure 3.5 A simple schematic eye. The eye adjusts the focal length of the lens to produce an image on the retina of objects at different distances. Consider for example, either one of the arrow objects. The location of the arrow's tip on the retina can be determined by tracing the principal ray through the center of the lens. The second principal ray enters the lens horizontally and passes through the focus on the far side of the lens before converging with the first principal ray. This second principal ray shows the required location of the focus in order that an image be formed on the retina.

explained using a principal ray analysis in a series of examples below. In the microscope and telescope, the eyepiece functions like a magnifying glass to magnify an intermediate image formed by the objective lens at the front of the device. So, we will start by considering the magnifying glass.

Example 3.2 Magnifying Glass The magnifying glass works by placing the object inside the focus of a single positive lens. When viewed from the right-hand side, the rays emerging from the lens look as if they're coming from an enlarged version of the object, a virtual image. This is illustrated in the figure herein.

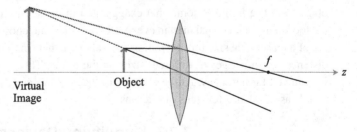

Two principal rays from the object are shown as solid lines. Their extension to the virtual image are dashed. To be convincing, this figure requires a clear understanding of why the virtual image is an image at all. So, we should remind ourselves that it's an image because each point on the virtual image radiates in all directions and the spatial relationship between points on the image is the same as for the object. The image is not distorted or mixed up somehow. To convince ourselves of this, we should make sure that the rays emanating from the magnifier do actually form a larger image on the retina than the image formed by viewing the object directly. This is shown in the figure below, where we've added a lens corresponding to the eye. The locations and size of the object and virtual image are preserved but we now treat the virtual image as if it was an actual physical object, ignoring the presence of the magnifying lens. (After all, the rays emerging from the magnifying lens are exactly the same as if they were emerging from a physical object as large as the virtual image sitting at the location of the virtual image.) We trace two principal rays from each object through the eye, solid lines for the original object, dashed for the virtual image.

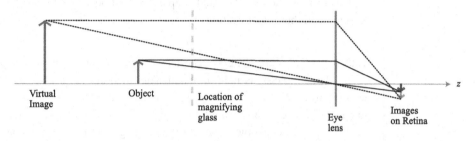

The top principal rays from the virtual image or the original image do not correspond to any actual rays entering the eye. The eye's pupil is so small that those rays would fall outside the pupil. However, the technique of principal rays still works. As we know, the size of the pupil only changes the brightness of the image on the retina; it doesn't change the image's size or location. Similarly, the larger the magnifying glass, the brighter the virtual image because as the solid angle presented by the magnifying glass to the object increases, it captures more rays. Regardless of brightness, the figure clearly shows that the rays corresponding to the virtual image throw a larger image onto the retina than the rays from the original object. In other words, the object has been magnified, as advertised.

Example 3.3 Microscope   Microscopes are basically an enhanced version of the magnifying glass. The trick is to add a second lens in front of the magnifying glass (now called the eyepiece). The second lens is known as the objective lens and its function is to create an intermediate image for the eyepiece to magnify. The intermediate image is larger than the original object, so the overall magnification is increased beyond what the eyepiece could provide by itself. The figure below demonstrates the configuration.

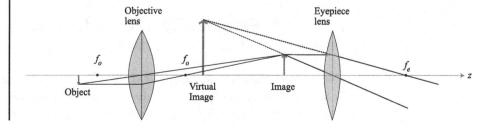

The focal length of the objective lens is f_o and the focal length of the eyepiece lens is f_e. In this figure, I put the object (far left) upside down so that I could simply reuse the figure from the magnifying glass, as-is. This was intended to emphasize the lack of mystery around the microscope. It really is just two magnifying stages strung together. The image formed by the objective lens is larger than the object and the virtual image generated by the eyepiece is larger still. Typical magnification values for an objective might be 10× or 40×. The eyepiece might contribute another 10× magnification, giving an overall magnification value of 100× or 400×. Obviously, the magnification of the arrangement shown in the image is chosen to be much lower (about 6×) so that object be visible and the images of reasonable size to fit on the page. Normally, the focal lengths of both the objective and eyepiece would be smaller than shown in the figure, leading to higher magnification.

Example 3.4 Telescope   The telescope is very similar in operation to a microscope. The emphasis is now on magnifying medium-sized and large things far away rather than small things close-up. To do that, we make the focal length of the objective longer and move it away from the eyepiece. The image formed by the objective lens is now smaller than the object. The virtual image generated by the eyepiece spans a much larger angle with respect to the viewer than the original object would when viewed with the naked eye. (For emphasis, I've again left the eyepiece unchanged from the case of the magnifying glass.) So the telescope has magnified the distant object by the ratio of the angles. It's also clear from the diagram why a telescope appears to "bring objects closer."

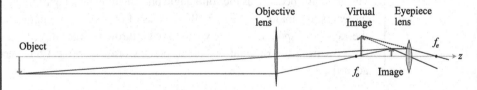

Once again, to keep the object and image sizes manageable, the telescope shown in this example doesn't have very high magnification and is set up to image an object unrealistically close to the telescope. A real telescope would have an eyepiece with much shorter focal length in order to magnify the image formed by the objective much more than shown here. That increases the angle subtended by the virtual image, increasing the overall magnification.

3.4 Paraxial Geometric Optics

If the angle, θ, that rays make with the optic axis is always small, we can make use of the fact that $\theta \approx \sin\theta \approx \tan\theta$. This is the same paraxial approximation we used in Chapter 1. It leads to simplification of the equations governing the behavior of light. For example,

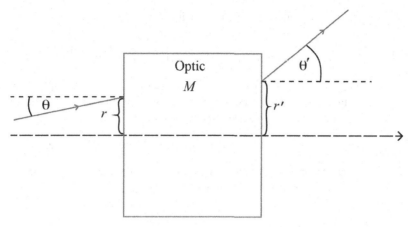

Figure 3.6 Schematic representation of an optic M.

under this approximation, Snell's law becomes $n_i\theta_i = n_t\theta_t$. In Chapter 4, we apply the same approximation to the case of physical optics where it leads to wonderful simplifications of otherwise intractable theory. Because optics is so simplified under the paraxial approximation, it often makes sense to treat a system paraxially even when the paraxial approximation is quite poor. Doing so, may allow you to quickly understand a complex system's behavior or get an idea of the effects of various design choices before doing a full simulation or building a prototype system.

For axisymmetric optical systems, the paraxial approximation also makes it easy to predict the effect of multiple optical elements. Each optical element is described by a 2×2 matrix

$$M = \begin{pmatrix} A & B \\ C & D \end{pmatrix} \tag{3.13}$$

known as the ray transfer matrix or more commonly the "ABCD matrix."[3] The A, B, C, D coefficients describe the action of the optic on a ray passing through the optic. Let r be the distance of a ray from the optic axis immediately before it encounters the optic M. Let r' be the distance of that same ray from the optic axis immediately after leaving the optic M. Similarly, let θ be the angle the ray makes with the optic axis immediately before M and let θ' be the angle immediately after M (see Figure 3.6). If the output height r' and output angle θ' are linearly related to r and θ, then we can write

$$r' = Ar + B\theta \tag{3.14}$$

$$\theta' = Cr + D\theta \tag{3.15}$$

---

[3] The treatment in this section uses the same notation as that of Fowles (1989). As usual, Siegman (1986) provides an excellent but more advanced treatment for readers wanting to extend the formalism to so-called ducts, which are optical elements with a quadratic, transverse index variation. This formalism also applies to physical optics and the extension to non-axisymmetric beams was recently provided by Magaña-Sandoval et al. (2019).

and these equations can then be written in matrix form as

$$\mathbf{r}' \equiv \begin{pmatrix} r' \\ \theta' \end{pmatrix} = \begin{pmatrix} A & B \\ C & D \end{pmatrix} \begin{pmatrix} r \\ \theta \end{pmatrix} \equiv M\mathbf{r}. \tag{3.16}$$

When multiple optics are placed in succession, we just multiply their ABCD matrixes. For example, with two optics, the ray $\mathbf{r}'$ exiting the first optic M_1 immediately enters the second optic M_2. The ray $\mathbf{r}''$ exiting the second optic is then just

$$\mathbf{r}'' = M_2\mathbf{r}' = M_2(M_1\mathbf{r}) = M_2M_1\mathbf{r}. \tag{3.17}$$

In other words, the matrix corresponding to the combined action of the two optics is just $M = M_2M_1$. For N of optics in succession then

$$M = M_N M_{N-1} \ldots M_1. \tag{3.18}$$

Note that the matrixes are multiplied in the *reverse* order that they are encountered by the ray. The action of an arbitrary combination of optics can therefore be reduced to the four components of a single 2×2 matrix!

Example 3.5 Free Space Propagation   If the "optic" is just empty space of length L, then the angle θ' of the beam exiting the region of empty space will be equal to the angle entering θ. The height r' of the beam exiting is just the height r of the beam entering plus an additional $L \tan \theta$ due to the propagation.

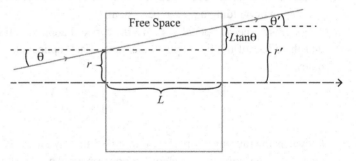

Since we are in the paraxial approximation $\tan \theta \approx \theta$, so

$$r' = r + L\theta \tag{3.19}$$
$$\theta' = \theta. \tag{3.20}$$

The *ABCD* matrix for free space propagation is therefore

$$M_L = \begin{pmatrix} 1 & L \\ 0 & 1 \end{pmatrix}. \tag{3.21}$$

In fact there is no difference between free space propagation and propagation within any dielectric from this point of view. As long as there is no *change* in refractive index, the matrix for free space propagation holds for propagation within any homogeneous dielectric. However, entering or exiting a dielectric from air involves a change of index and therefore additional ABCD matrixes. See Example 3.7 for an application.

Example 3.6 Propagation through a Thin Lens   The action of a thin convex lens in the paraxial approximation is to turn an incoming ray through an angle $\Delta\theta$ toward the optic axis. The angle through which the ray is turned is proportional to the height at which the ray intercepts the lens. $\Delta\theta = r/f$, where f is the focal length of the lens.

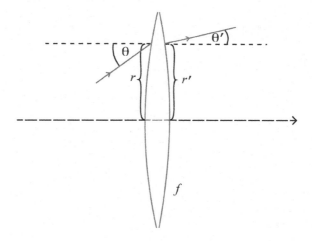

In other words,

$$r' = r \tag{3.22}$$

$$\theta' = \theta - r/f, \tag{3.23}$$

where the negative sign is due to the fact that for positive f, the ray is turned toward the optic axis. So the $ABCD$ matrix for a thin lens is

$$M_f = \begin{pmatrix} 1 & 0 \\ -1/f & 1 \end{pmatrix}. \tag{3.24}$$

Example 3.7 ABCD Matrix for a Window   A window (in air) with index n, of thickness L and with parallel plane surfaces and perpendicular to the optic axis can be described by the ABCD matrix

$$M_w = \begin{pmatrix} 1 & 0 \\ 0 & n \end{pmatrix} \begin{pmatrix} 1 & L \\ 0 & 1 \end{pmatrix} \begin{pmatrix} 1 & 0 \\ 0 & \frac{1}{n} \end{pmatrix} \tag{3.25}$$

$$= \begin{pmatrix} 1 & \frac{L}{n} \\ 0 & 1 \end{pmatrix}. \tag{3.26}$$

The left and right factors in Eq. (3.25) are just the matrixes for a plane dielectric boundary when exiting and entering the window respectively. The middle matrix in the product is the matrix for free space propagation. Note that the second row of the resulting matrix implies that rays exit the window with the same angle as they entered.

Table 3.1 ABCD matrixes for simple optics.

Description		M	Comments
Free propagation		$\begin{pmatrix} 1 & L \\ 0 & 1 \end{pmatrix}$	Applies to propagation in vacuum or any linear dielectric.
Thin lens		$\begin{pmatrix} 1 & 0 \\ -\frac{1}{f} & 1 \end{pmatrix}$	f is positive for convex lenses.
Spherical mirror		$\begin{pmatrix} 1 & 0 \\ -\frac{2}{R} & 1 \end{pmatrix}$	R is positive if the mirror is concave toward the incoming rays, as shown.
Plane dielectric boundary		$\begin{pmatrix} 1 & 0 \\ 0 & \frac{n}{n'} \end{pmatrix}$	
Spherical dielectric boundary		$\begin{pmatrix} 1 & 0 \\ \frac{1-n/n'}{R} & \frac{n}{n'} \end{pmatrix}$	R is positive if the boundary is concave as shown.

3.4.1 Effect of an Optic on Rays Emanating from a Point

Consider the case of rays emanating from a point on the optic axis and passing through an optic M. How is the rate of divergence of rays entering the optic related to the rate of divergence of rays exiting the optic? To answer this, we consider the radii of curvature of "wavefronts" perpendicular to the emanating rays as shown in Figure 3.7. (Of course these aren't actual wavefronts since the rays are assumed to be incoherent; they are just circular contours running perpendicular to the rays.)

For the rays on the left $R = r/\theta$ and on the right $R' = r'/\theta' = (Ar + B\theta)/(Cr + D\theta)$. Dividing the numerator and the denominator by θ, gives

$$R' = \frac{AR + B}{CR + D}. \tag{3.27}$$

The rate of divergence of neighboring rays characterized as the radius of curvature of perpendicular contours ("wavefronts") immediately before and after the optic, R and R' respectively. In this sketch, R' is shown less than zero, while R is shown larger than zero, but any combination is possible. No assumption is made about the sign of R or R' in the result of Eq. (3.27).

This equation for the effect of an optic on the rays emanating from a point source will be very important when, in Section 4.1, we extend this method to the wavefronts of a Gaussian beam.

Exercises

3.1 A single positive lens can be used as a magnifying glass. Using principal rays, show that a single negative (concave) lens is a reducer. In other words, things should look smaller when viewed through a negative lens.

3.2 Consider a telescope consisting of two positive lenses, known as a Newtonian telescope. A collimated beam is a bundle of parallel rays, such as the output of a laser or rays coming from an object at infinity. Using principal rays, show that if a collimated beam enters the objective and exits the eyepiece as a collimated beam, then the distance between the lenses should be equal to the sum of the focal lengths. Also show that the width of the collimated beam is changed by the ratio of the focal lengths. Do the same for a telescope consisting of a positive and negative lens, known as a Galilean telescope.

3.3 A Galilean telescope consists of a divergent lens with focal length $-f_1$ separated by a distance L from a convergent lens with focal length f_2. Assume $f_1, f_2 > 0$ and $f_2 - f_1 = L$. Find the ABCD matrix for a Galilean telescope.

3.4 Show that the ABCD matrix for a planar dielectric interface is $\begin{pmatrix} 1 & 0 \\ 0 & \frac{n}{n'} \end{pmatrix}$ as shown in Table 3.1.

3.5 Consider a ray passing through a spherical interface between two dielectrics. The incident medium has index of refraction n and the transmitted medium has index n'.

If the radius of curvature of the interface is R (with $R > 0$ corresponding to the interface curving toward the incoming ray). Show that the ABCD matrix corresponding to the interface is

$$M_{\text{sph}} = \begin{pmatrix} 1 & 0 \\ \frac{n'-n}{n'R} & \frac{n}{n'} \end{pmatrix}$$

You may find the construction below useful.

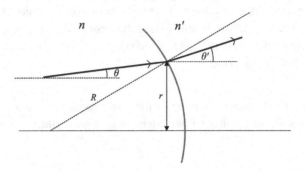

3.6 Use Snell's law and the result of Exercise 3.5 to obtain the ABCD matrix of a thick plano-convex lens.

3.7 Derive the expression given in Eq. (3.12) for the magnification $\frac{h'}{h}$ of a single lens imaging system. Use the notation of the diagram below. Rays are indicated (as usual) by the solid lines connecting the tips of the object and image arrows.

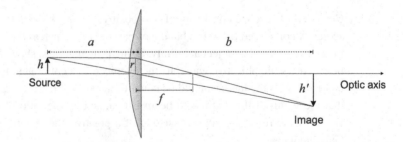

3.8 Consider a point source of spherical waves. Use Eq. (3.27) to find the radius of curvature of a mirror that refocuses the light to the point of emission. Does the result make sense?

3.9 Show that the focal length of a concave spherical mirror is half its radius of curvature.

3.10 Calculate an exact expression for the relationship between the incident angle θ_1 and the exit angle θ_3 for a ray passing through an equilateral prism with index of refraction $n > 1$. (An equilateral prism is a prism of triangular cross section with sides of equal length). See the figure below.

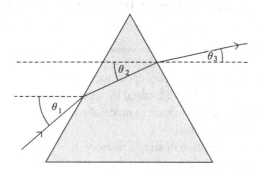

Note that the difference between the exit angle and incident angle is the same as the angle through which the prism turns the light, $\Delta\theta = \theta_3 - \theta_1$. From the above mentioned figure, you can see that $\Delta\theta$ will be negative, indicating that the beam turns clockwise. For simplicity, consider the case when the incident and exit angles are equal in magnitude but have opposite signs. The ray inside the prism is then traveling parallel to the base of the prism. Expand $\Delta\theta$ in terms of small changes in the index of refraction, Δn around $n = n_0$, where n_0 is the index at some reference wavelength λ_0. Take λ_0 to be some wavelength near the middle of the visible range. n_0 is then the index of refraction of some common glass at that wavelength. You'll need to look these quantities up for some glass of your choice. By also looking up the value of $\frac{dn}{d\lambda}\big|_{\lambda_0}$ for your choice of glass, estimate how far from the prism you would have to be for the width of the rainbow from a narrow collimated beam of sunlight to be one centimeter wide.

You should find that the linear term in the expansion is nonzero. Based on this result, and assuming that any dependence of n on wavelength is also roughly linear for wavelengths within the visible region (a fairly good approximation) do you expect that all colors (wavelengths) are about equally represented in the rainbow?

3.11 (Computer problem) Apply Snell's law twice to plot the path of twenty or so parallel rays incident on a spherical plano-convex lens. The rays encounter the flat side of the lens first.

3.12 (Computer problem) Consider parallel rays reflecting from the inside of half a circular hoop lying in the plane of the rays. Plot the path of forty or so equally spaced parallel rays incident on the hoop and covering the full diameter of the hoop. To prevent the figure from becoming too busy, only plot the rays from the point of reflection onward. In other words, don't show the incoming rays. The shape of the caustic that emerges is a nephroid and is sometimes seen on the surface of a cup of tea or coffee when drunk outside on a sunny day. (In optics, a "caustic" is an envelope of light rays forming a particularly bright curve or region.)

3.5 Experiment: Imaging Optics

Objectives

1 Gain a nuanced understanding of the concept of focal length.

2 Observe the detailed behavior of light focused by spherical lenses.

3 Write a Monte Carlo ray tracing program to model the region of a focus.

4 Measure the curvature of the cornea.

5 Build and understand the operation of simple telescopes.

6 Observe and analyze chromatic aberration.

7 Maximize the image sharpness in a pinhole camera and come up with a theory explaining the result.

Equipment Needed

- Three plano convex lenses of varying focal lengths, 50–1, 000 mm.
- One plano convex lens of fairly short focal length, 100–300 mm, having diameter greater than 250 mm.
- One opaque marble.
- Ruler graduated more finely than 1 mm, if possible.
- Several neutral density filters or several small sheets of unexposed, processed film.
- Lens tubes that can accommodate a lens on one end and slide/screw into one another. If commercially produced tubes are not available, these can be made from stiff cardboard or aluminum tubing.
- Large (10–50 cm diameter) spherical mirror with focal length 10–100 cm.
- Incandescent lamp with a clear glass bulb such as a halogen lamp.
- Tin foil and a sharp sewing needle, or a series of professionally prepared pinholes of various diameters between 5 and 100 µm.
- Medium-sized cardboard box.
- A distant, bright point-like or semi-point-like light source such as the sun, full moon, a street light, and so forth.
- Non-flammable screen, such as a piece of sheet-metal spray-painted matte white.

> *Looking directly at the sun is harmful to your eyes. Be especially careful with any lenses you may be using.* **Under no circumstance should you look at the sun through a lens or any optical device not explicitly designed for the purpose.** *Doing so is very likely to lead to vision loss.*

Focal Length

Verify the focal length of three convex lenses whose nominal focal length is known. Do this by imaging (focusing) a distant light source, like the sun or the full moon, onto a non-flammable projection screen. Make sure the lens is perpendicular to the entering rays so that you don't induce aberrations in the focused image. Choose lenses with a fairly wide range of focal lengths. Is the distance between the lens and the image a good measure of the focal length? Does the specified focal length always agree with your measurement? Are there any effects of chromatic aberration?

> *Recommendations*: Only remove one lens from the laboratory's lens collection at a time and return that lens before taking another. Lenses are typically somewhat small, fiddly, and expensive, so be very careful not to drop the lens whose focal length you are measuring.

The Neighborhood of the Focus

Obtain a large diameter $d > 2.5$ cm plano-convex lens, with a fairly short focal length $10 < f < 30$ cm. Measure/verify its focal length by imaging the distant light source (sun, moon, etc.). In order to investigate spherical aberrations, you will photograph the image of a distant source when it's in focus and also just out of focus. Take photographs of the image in the following configurations:

- Lens' convex surface faces the image and the distance from the lens to the projection screen is just a little shorter than the measured focal length. The lens should be just a few percent of a focal length inside the distance at which the image is "in focus."
- Lens' convex surface faces the image and the distance from the lens to the projection screen is equal to the measured focal length.
- Lens' convex surface faces the image and the distance from the lens to the projection screen is a little larger than the measured focal length. The lens should be just a few percent of a focal length beyond the distance at which the image is "in focus."

Import the photographs into Matlab or Python (etc.) and produce a single plot showing the irradiance as a function of distance from the optic axis in each of the cases mentioned earlier. Compare the results to a Monte Carlo simulation you perform that propagates a random uniform distribution of parallel rays through a lens to the appropriate position near the focus. The steps of the simulation are:

- Define a ray traveling normal to the lens that enters the lens at some *random* position on the lens.
- Use Snell's law at the curved surface to propagate the ray through the lens. Note that the curved surface is spherical.
- Find the position at which the ray hits the screen. On a plot representing the screen, draw a single dot at that location.
- Repeat the above steps until you have a sufficiently dense image of the light distribution on the screen.

To produce a plot of the irradiance as a function of distance from the optic axis, you could have your code count the number of rays passing through annuli of increasing diameter. Figure 3.8, produced by tracing a selected set of rays through a plano-convex lens may help you to visualize what to expect.

Recommendations: If you are using the sun, **do not place the camera directly into the focused light**. You are likely to destroy it. Instead, take a photo of the light on the screen. Even with dimmer sources than the sun, the main difficulty will be obtaining a photograph that does not saturate the CCD or CMOS array in the camera. You may need to adjust any aperture and exposure controls your camera may have. In addition, it may be necessary to place very dark glass (like a neutral density filter) in front of the camera lens to prevent saturation. If, during analysis, you notice that your irradiance curves are flat-topped, the camera was probably saturated and you will likely need to redo this measurement. If you choose to use a phone camera, you will need access to exposure control settings that may require additional apps.

Make sure you orient the lens so that the flat side faces the sun. That surface doesn't participate in the refraction since the rays pass through at normal incidence. This makes the simulation easier.

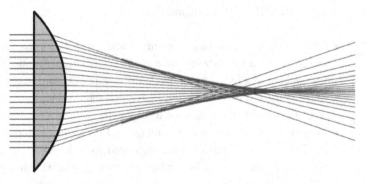

Figure 3.8   The paths taken by parallel rays entering a plano-convex spherical lens at different heights show that even in the geometric optics approximation, a spherical lens does not produce a "perfect focus." The image shows individual rays being traced but also highlights the envelope of the rays where light is particularly concentrated (the "caustic").

The Thin Lens Equation

Using the lens from the previous section, image an extended light source such as a bulb with a filament onto a screen or a wall. (The filament is the object.) Quantitatively verify the thin lens equation, Eq. (3.11). Does the thin lens equation hold within the uncertainties you have assigned? Note that the thin lens equation is symmetrical in a and b. In other words, it would seem that given a particular distance $a + b$ between the source (object) and the screen (image), there are two different lens locations that will generate an image. Verify this. If $d = a + b$ is the distance between the source and screen, show that for any given lens with focal length f, no image can be produced when $d < d_{min}$ for some minimum source-to-screen distance, d_{min}. Verify the result of your derivation by showing that when the object is moved closer to the screen than d_{min}, the lens in fact fails to produce a sharp image, regardless of where it is placed.

Recommendations: In the absence of an optical rail assembly, a good way to find the appropriate lens location for a fixed object-to-image distance is to tape a meter stick onto a table with one end of the meter stick against the wall. Place the bulb some distance away from the wall along the meter stick. Holding the lens by the edges (and using gloves) run the lens along the meter stick until the filament is imaged on the wall.

Reflection From a Convex Spherical Surface

Place a glass marble, preferably the solid-color kind, on a table underneath a ceiling-mounted fluorescent light. Using a finely graduated ruler, or a pair of calipers, measure the size of the reflection of the light fixture in the marble. Also, measure the size of the actual fixture and the distance from the fixture to the center of the marble. Show that the radius of the marble is given by

$$R \approx \frac{d/2}{\tan\left[\frac{1}{2}\arctan\left(\frac{D}{2L}\right)\right]}, \qquad (3.28)$$

where the variables refer to Figure 3.9. Note that the size of the marble relative to the fixture is greatly exagerated in the diagram. Therefore, the approximation that $\varepsilon \ll \theta$ is far better than it looks in the diagram. For your estimate of the uncertainty, note that your largest source of uncertainty is likely to be the measurement of d.

Since the eyeball is also a somewhat reflective sphere, the same principle can be used to measure the radius of curvature of the cornea. Have your lab partner lie on the floor or on a table under a light fixture so that the reflection of the light fixture is clearly visible from directly above your partner. Applying a ruler or calipers directly to the eyeball risks injuring the eye. Instead, place a finely graduated ruler on your partner's forehead or cheek and take a photograph of your partner's face from directly above and about half a meter away. Make sure the photo captures both the reflection of the fixture and the ruler. Instead of measuring the size of the reflection directly as you did in the last part, you now have a calibrated photograph on which you can measure the size of the reflection. I re-iterate, don't try to measure the size of the reflection of the fixture directly on the eye! Use the photograph instead. All other steps are the same as for the marble. A careful set of measurements with this technique may even be able to detect astigmatism due to a slight ellipticity of the cornea. In that case, the measured radius of curvature will be different when measured in two orthogonal directions.

Recommendations: To measure the distance from your partner's eye to the fixture, measure the height of your partner's eye above the table or floor and then measure the distance between the table or floor and the fixture.

Refractive Telescopes

Build a Keplerian telescope. A Keplerian telescope consists of two positive (convex) lenses with different focal lengths. The focal length of the objective should be longer than that of the eyepiece lens. (For example: f = 400 mm for the objective versus f = 50 mm for the eyepiece.) The magnification is given by the ratio of the focal lengths. Note that a

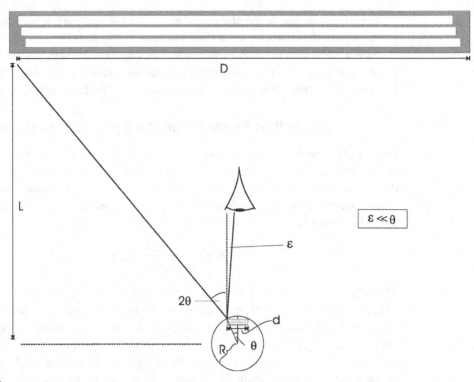

Figure 3.9 Schematic of a spherical marble reflecting the outermost ray from a light fixture into the eye. The condition of specular reflection off a spherical surface is captured by the statement that the angle subtended by the incident ray and the vertical is twice the angle from the vertical to point of reflection on the marble, 2θ and θ respectively. We make the approximation that the size of the image is small compared with the distance of the viewer from the marble. In that case, all the rays reflecting from the marble and imaged by the eye can be considered to be vertical after leaving the marble.

telescope is focused on objects at infinity when parallel rays entering the objective lens exit the eyepiece also as parallel rays. The lenses should be separated by the sum of their focal lengths.

Using the same objective lens, but exchanging the eyepiece lens for a negative (concave) lens, also build a Galilean telescope. (Note that the lens separation is still the sum of the focal lengths, but now one of the focal lengths is negative, so the lens tube will need to be shortened.) By sketching the path of light rays through the device, explain why the image is inverted in one telescope but not in the other. By observing an object with sharp changes in contrast (e.g., the black lines on a resolution chart) see if you see any chromatic aberration. Chromatic aberration occurs when the refractive power of the lenses is wavelength dependent. Red colors may be correctly imaged when blue colors are out of focus and vice versa. In objects that contain a mixture of wavelengths (most objects) a colored halo appears around sharp contrast changes such as the edge of a roof against the sky. Describe this halo in detail in your lab book. Is there a pattern to the order of the colors in this halo? Does this depend on how you focus the telescope?

> *Recommendations*: Use tube assemblies that slide or screw together in such a way as to make the distance between the lenses easily adjustable. To prevent your eyelashes or glasses from touching the eyepiece lens, mount the lens slightly inside the end of the tube or attach an additional short piece of tubing between the eye and the eyepiece lens. Calculate the required length of the lens tubes before choosing the parts.

Reflecting Telescopes

Build a reflecting telescope with a eyepiece and a concave mirror. Why does this telescope not have any chromatic aberration? Although we have replaced the objective lens with a mirror, the eyepiece lens' refractive power still depends on wavelength. Use a ray diagram to explain why chromatic aberration is much lower than in the refractive telescope.

> *Recommendations*: Don't bother with a flat mirror to pick off the beam from the concave mirror. Instead, operate the instrument with the eyepiece slightly off axis. It is sufficient for one person to hold and point the mirror. The viewer can then hold the eyepiece in the appropriate viewing position. No mechanical structure is needed.

Small Apertures

Use a sewing needle to make a tiny aperture in a piece of foil. If you have access to a series of professionally produced pinhole apertures in various sizes, you could use one of those instead. Behind your pinhole aperture, place an incandescent lamp with a clear (unfrosted) bulb. On the other side of the aperture, place a screen. This is a pinhole camera but with the screen replacing the film. If the pinhole aperture is small enough, you should see a dim image of the lamp's filament on the screen. Experiment with different sized pinholes to obtain the size that produces the sharpest image. Note that after a certain point, reducing the size of the pinhole leads to a blurrier image. This is contrary to what you might expect from geometric optics alone. What two effects causing reduced image sharpness are approximately equally large for the aperture leading to the sharpest image?

> *Recommendations*: In order that your screen be in shadow, you may want to tape the foil containing your pinhole across a hole in the end of a medium-sized box. Put the light source outside the box and almost touching the foil. The image projected on the inside of the box is likely to be quite dim so tape a piece of white paper to the inside of the box as a screen. (It's also helpful to turn the box onto its side so that the open top of the box faces you while light enters from the end of the box to your left or right. One of the flaps can serve to shield you from the light source.) Another possibility is to keep the box completely closed and make a viewing hole with a tube for the eye. You can also incorporate an input tube for the pinhole. Doing so allows the foil to be wrapped over the end of the input tube and pinholes are then more easily changed. One way to vary the size of a pinhole made with a needle is to place the aluminum foil on surfaces of various hardness.

Ideas for Further Investigation

A simple microscope can be built from two lenses. Build one and test it. How could it be improved?

Repeat the investigation of the neighborhood of a focus with the lens flipped over so that it is oriented with the convex side toward the light source. Compare the aberrations generated by the two orientations of the lens.

Your Monte Carlo ray tracing program could be modified to accommodate chromatic aberration. Images of the sun or moon could be obtained in color and the result compared to the modified program. To implement the wavelength dependence of the index of refraction of glass, either interpolate tabulated values of n versus λ or use Sellmeier equations.

4 Physical Optics

Now we relax the assumption made in geometric optics that the wavelength tends to zero. We also allow coherent fields. The variety and sheer number of phenomena now encompassed by the theory is so large that we can't hope to cover them in a single book, let alone a single chapter. So, we must choose a way forward that serves our immediate purpose. Almost every optics laboratory uses lasers. Understanding their output beams is crucial to being able to work in an optics lab at all. We pick up where we left off in Chapter 1 and specialize the paraxial approximation to Huygens' integral further to allow us to describe laser beams.[1] The method is then taken to its logical conclusion in the application of the Fourier transform as a beam propagation method. The method is not always expounded in much detail in optics books[2] but it's such a useful and general way of propagating laser beams in a realistic optical system, especially where the finite diameter or surface imperfections of the components has an effect, that I think it's worth learning to do.

In order to emphasize the fundamental role of diffraction in physical optics, the experiment at the end of this chapter concentrates on diffraction and on one of the major applications of diffraction: spectroscopy. The theory leading to Gaussian beams and to the Fourier transform method may seem rather "heavy" in relation to the experiment, especially when we can already use Huygens' integral to calculate diffraction patterns numerically. However, the theoretical material in this chapter is also needed for later chapters. The experiment in Chapter 6 relies heavily on the content from this chapter and the discussion of cavities in Chapter 7 benefits from having read this chapter first. I urge you to take the time to study it.

4.1 Paraxial Physical Optics

We limit ourselves to the paraxial approximation where wavefront normals make small angles with the optic axis. So, beams are not diverging or converging quickly. The paraxial approximation is appropriate for most systems using lasers as a source of light. As we shall see, all of the machinery of paraxial geometric optics carries over to paraxial physical optics with only a "small" modification to the concept of the beam radius of curvature.

---

[1] Section 4.1.1 outlines the theory in a way that uses the same notation and general approach as numerous other texts, for example, Svelto (2010) or Pedrotti (2007). To get more details than provided here, I also recommend Siegman (1986).

[2] One exception is Hecht (2017) who provides a full chapter on Fourier optics.

4.1.1 Self-Similar Beams Propagating in Free Space: TEM_{mn} Modes

Anyone who has played with a flashlight can attest to the fact that the beam shone on a nearby wall looks fairly different than when it is shone on a faraway wall. And the difference is more than just an overall change in brightness. Usually, the beam's shape is different, perhaps having a halo on the faraway wall where none is apparent on the nearby wall. In other words, the beam seems to change its transverse profile in a nontrivial way as it travels. An interesting question one might ask about light propagation is whether there are any beams that *retain* their transverse profile as they propagate. Quantitatively, we are asking whether there are any beam shapes $u(x, y, z_1)$ that retain their *functional form* after propagation by Eq. (1.31). The answer to the question is yes! The so-called TEM_{mn} free-space modes of the electromagnetic field retain their functional form after propagation. "TEM" stands for "Transverse Electric and Magnetic" and m and n are nonnegative integers labeling the different possible modes. The output of most lasers is a TEM_{00} mode, often called a "Gaussian beam," so propagation of free-space modes of the electromagnetic field holds more than academic interest.

While TEM_{mn} modes retain their functional form during propagation, the overall amplitude and phase of the beam does change. The size of the beam and wavefront curvature also evolves. To see how, we need to consider the propagation of the field amplitude u corresponding to a TEM_{mn} mode. In some source plane located at $z = z_1$, the complex field amplitude u of a TEM_{mn} mode is

$$u(x, y, z_1) \equiv u_0 H_m\left(\tfrac{\sqrt{2}x}{w_1}\right) H_n\left(\tfrac{\sqrt{2}y}{w_1}\right) e^{-ik\frac{x^2+y^2}{2q_1}}, \tag{4.1}$$

where $u_0 \in \mathbb{C}$ is an overall constant, the "beam radius" w_1 is the distance from the optic axis at which the field amplitude of a TEM_{00} mode has fallen in magnitude to $\frac{1}{e}$ of its maximum value. The quantity $q_1 \in \mathbb{C}$ is known as the "complex radius of curvature," or "complex beam parameter." It specifies both the wavefront curvature and the size of the beam in the source plane. The complex beam parameter is defined in terms of the beam radius w and the wavefront radius of curvature R according to

$$\frac{1}{q} \equiv \frac{1}{R} - i\frac{\lambda}{\pi w^2}. \tag{4.2}$$

After propagating a distance L in free space according to Eq. (1.31), it can be shown that the complex field amplitude becomes

$$u(x, y, z) = \zeta u_0 H_m\left(\tfrac{\sqrt{2}x}{w}\right) H_n\left(\tfrac{\sqrt{2}y}{w}\right) e^{-ik\frac{x^2+y^2}{2q}}, \tag{4.3}$$

where q is the new complex beam parameter given by

$$q = q_1 + L. \tag{4.4}$$

The wavefront radius of curvature R and beam radius w after propagation are now obtained from q. Since Eq. (4.2) is a complex-valued equation, the real and imaginary parts are independent equations allowing us to solve it for both w and R. The complex factor ζ out

front in Eq. (4.3) indicates that the overall amplitude and phase offset of the beam have
changed. It can be shown that the propagation integral leads to

$$\zeta \propto \left[\frac{q_1}{q_1 + L}\right]^{1+m+n}.\tag{4.5}$$

So, if the source field is a TEM$_{mn}$ mode, we can propagate it in free space using the
equations mentioned herein without needing to perform the integral 1.31 explicitly. Really,
we just need Eqs. (4.4) and (4.5) to know how the beam has changed. This is an enormous
simplification and allows us to propagate laser beams much more easily than we otherwise
could.

We'd like to investigate the form and behavior of the TEM$_{mn}$ modes further. For that, it's
best to work in a coordinate system where the algebra is as simple as possible. We consider
a TEM$_{mn}$ mode propagating in free space, traveling in the $+\hat{z}$ direction along the z–axis, and
having *planar* wavefronts at $z = 0$. (In other words $R \to \infty$ as $z \to 0$.) The source plane is
the xy–plane ($z = 0$). In the source plane, Eq. (4.2) gives $q_1 \equiv i\pi w_0^2/\lambda$ since $R \to \infty$. The
beam radius in the source plane is w_0.[3] To find the beam radius w and radius of curvature
R at some downstream location z, we apply the propagation formula, Eq. (4.4), replacing
L with z. Solving Eq. (4.2) for R and w, we get (with some algebra)

$$R = \frac{1}{\mathrm{Re}\{q^{-1}\}} = z\left[1 + \left(\frac{z_0}{z}\right)^2\right]$$

$$w = \left[-\frac{\lambda}{\pi\,\mathrm{Im}\{q^{-1}\}}\right]^{\frac{1}{2}} = w_0\left[1 + \left(\frac{z}{z_0}\right)^2\right]^{\frac{1}{2}}$$

where

$$z_0 \equiv \frac{\pi w_0^2}{\lambda}.\tag{4.6}$$

The quantity z_0 is known as the Rayleigh range of the beam. At one Rayleigh range from
the waist, the beam irradiance has dropped by a factor of two compared with the irradiance
at the waist. The Rayleigh range also corresponds roughly to the distance beyond which
the wavefronts propagate like spherical waves and the wavefront radius of curvature, R,
becomes approximately equal to z. Clearly, w increases for both positive and negative z, so
w_0 is also the minimum beam radius, known as the "beam waist."

These equations, which give the wavefront radius of curvature and beam radius as a
function of position along the optic axis z, are so important that I'll emphasize them with a
box.

$$R(z) = z\left[1 + \left(\frac{z_0}{z}\right)^2\right]\tag{4.7a}$$

$$w(z) = w_0\left[1 + \left(\frac{z}{z_0}\right)^2\right]^{\frac{1}{2}}\tag{4.7b}$$

---

[3] We reserve the subscript "0" for quantities associated with flat wavefronts, $R \to \infty$. As we shall see, flat
wavefronts occur at the minimum beam width known as the waist. So, w_0 is the radius of the beam waist.

As noted earlier, the phase offset also evolves. The phase offset is given by the argument of the complex amplitude factor in Eq. (4.5). It can be rewritten as

$$\zeta \propto \left(\frac{1}{1 - i\eta}\right)^{1+n+m} \qquad \text{where} \quad \eta = \frac{z}{z_0},$$

where we've replaced L with z and use the fact that the waist is at $z = 0$. Applying the relationship $z = |z| \exp\left[i \arg(z)\right]$, we get

$$\left(\frac{1}{1 - i\eta}\right)^s = \left\{\left(\frac{1}{\sqrt{1 + \eta^2}}\right) \exp\left[-i \arg(1 - i\eta)\right]\right\}^s = \left(\frac{1}{\sqrt{1 + \eta^2}}\right)^s \exp\left[is \tan^{-1} \eta\right], \quad (4.8)$$

where $s = 1 + n + m$. Then using Eq. (4.7b) to rewrite $\sqrt{1 + \eta^2}$ as $\frac{w}{w_0}$ and normalizing so that the power in the propagated beam Eq. (4.3) is equal to the power in the original beam Eq. (4.1), we find

$$\zeta(z) = \left(\frac{w_0}{w}\right) \exp(i\phi_0), \qquad (4.9)$$

where

$$\boxed{\phi_0(z) = (1 + n + m) \tan^{-1}\left(\frac{z}{z_0}\right)} \qquad (4.10)$$

The phase ϕ_0 is known as the Gouy phase of the beam and is associated with the change from leftward curving wavefronts to rightward curving wavefronts. The total Gouy phase change as a beam approaches a focus from $-\infty$, passes through the focus, and propagates to $+\infty$ is

$$\Delta\phi_0 = (1 + m + n)\pi. \qquad (4.11)$$

So we see that a TEM$_{00}$ mode will gain a maximum phase offset of π as it passes through a focus. Higher order modes $m, n > 0$ gain more.

Equations (4.7) and (4.10) encapsulate the free-space propagation behavior of the TEM$_{mn}$ modes. Substituting Eq. (4.2) into Eq. (4.3) leads to the explicit expression for the scalar field amplitude

$$\boxed{\begin{aligned} \tilde{E}(x, y, z) &= u(x, y, z)e^{-ikz} \\ &= u_0 \frac{w_0}{w(z)} H_m\left(\frac{\sqrt{2}x}{w(z)}\right) H_n\left(\frac{\sqrt{2}y}{w(z)}\right) e^{-\frac{x^2+y^2}{w^2(z)}} e^{-ik\left[z + \frac{x^2+y^2}{2R(z)}\right] + i\phi_0} \end{aligned}}$$

$$(4.12)$$
$$(4.13)$$

As in Eqs. (4.7) and (4.10), we express the parameters, $w(z)$, $R(z)$, and $\phi_0(z)$, as functions of position z, assuming the waist is at $z = 0$. The squared magnitude of this field amplitude profile is the irradiance profile of the beam

$$\boxed{I(x, y, z) = I_0 \frac{w_0^2}{w^2(z)} H_m^2\left(\frac{\sqrt{2}x}{w(z)}\right) H_n^2\left(\frac{\sqrt{2}y}{w(z)}\right) e^{-\frac{x^2+y^2}{\left[w(z)/\sqrt{2}\right]^2}}} \qquad (4.14)$$

where $I_0 \equiv I(0, 0, 0)$ is the irradiance on the optic axis at the beam waist. The transverse irradiance profile of such beams is illustrated in Figure 4.1. As mentioned earlier, the lowest

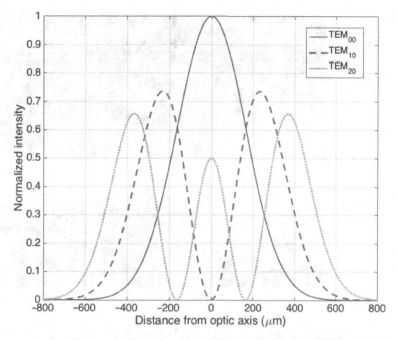

Figure 4.1 The transverse irradiance profiles of modes TEM$_{00}$, TEM$_{10}$, and TEM$_{20}$ along the
x-axis. The total power in the three beams is the same.

order mode, TEM$_{00}$, is the type of beam produced by most lasers and is therefore the
most common type of beam encountered in a lab. The shape of such a Gaussian beam
propagating in free space is illustrated in Figure 4.2.

One aspect of note is the relationship between the waist size and the angle of divergence
of the beam. (See Figure 4.3). Differentiating Eq. (4.7b) with respect to z and taking $z \to \infty$
leads to

$$\theta_0 \approx \left(\frac{\lambda}{\pi}\right) w_0^{-1}, \tag{4.15}$$

where θ_0 is the divergence (or convergence) angle as the beam leaves (approaches) the
waist, w_0. This means that to get a small focus (small beam waist), one must use a short
focal length lens, also known as a "fast" lens. In practice, one needs to use a lens with a
parabolic profile rather than a spherical profile if one wants to get the smallest possible focus. Such "aspheric" lenses can achieve a diffraction limited focus, even for large incident
beams where the spherical aberration from normal lenses would spread out the focus.

The fact that small waists correspond to rapid spreading can be understood as a consequence of the Heisenberg Uncertainty Principle. Photons passing through the waist have
a spread in their transverse momenta corresponding the transverse momentum uncertainty
Δp_y. Those same photons have a transverse position uncertainty Δy, corresponding to the
waist size. For a Gaussian beam, the uncertainty product is minimized: $\Delta y \Delta p_y = \hbar/2$.
Therefore, smaller waist sizes must lead to photons with a larger spread of transverse momenta. As these photons leave the waist, the beam spreads out more quickly due to the
larger momentum spread.

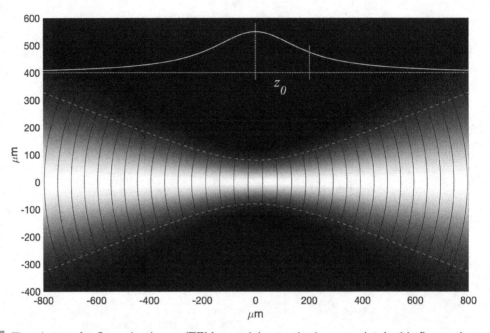

Figure 4.2 The shape of a Gaussian beam (TEM_{00} mode) near the beam waist. In this figure, the irradiance in each transverse plane (perpendicular to the optic axis) has been scaled so that the irradiance on the optic axis is always unity. This was done so that the transverse profile can more easily be seen; otherwise the region of the focus would show up very brightly while the remainder of the beam would be too dim to register on the graph. The dashed curves show the transverse distance at which the field *amplitude* has fallen to $1/e$ of its value on the optic axis (the *irradiance* has fallen to $1/e^2$ of the irradiance on the optic axis). The solid curve on a separate set of axes near the top of the figure shows the actual irradiance of the light on the optic axis relative to the irradiance at the waist. The irradiance on the optic axis falls by a factor of two relative to the waist once the Rayleigh range z_0 is reached. The black curves in the beam show the locations where the field amplitude is zero. This happens twice per wavelength. In this simulation, the wavelength was chosen to be $100 \, \mu\text{m}$.

4.1.2 Propagating a Beam through an ABCD Optic in the Paraxial Approximation: A Fourier Transform

In the last section, we discussed beams propagating in free space. Now we will propagate a beam through an arbitrary optic described by some ABCD matrix. If the beam incident on an optic has complex field amplitude $u(x, y, z_1)$ as before, then Eq. (1.31) can be generalized (see e.g., Siegman (1986)) to give the beam immediately after the optic as

$$u(x, y, z) = \frac{i}{B\lambda} \iint\limits_{\mathbb{R}^2} u(x', y', z_1) \, e^{-ik\frac{A(x'^2+y'^2)+D(x^2+y^2)-2xx'-2yy'}{2B}} \, \mathrm{d}x' \, \mathrm{d}y'. \quad (4.16)$$

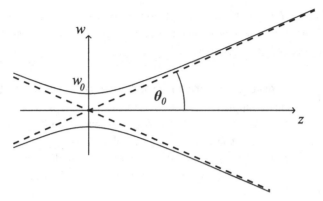

Figure 4.3 At distances from the waist $z \gg z_0$, the beam radius, w, grows linearly. The smaller the waist, the faster the beam radius grows.

If $u(x, y, z_1)$ corresponds to a TEM$_{mn}$ mode with complex beam radius q_1, the propagated complex beam radius q is

$$q = \frac{Aq_1 + B}{Cq_1 + D} \qquad (4.17)$$

For free space propagation ($A = 1$, $B = L$, $C = 0$, $D = 1$), this equation reverts to Eq. (4.4) as expected. It's worth pausing to emphasize the utility of Eq. (4.17). This formula allows us to propagate any TEM$_{mn}$ mode through any optics chain described by ABCD matrixes. That's really useful! We can finally deal with laser beams properly. After propagation, we can find the new beam radius and wavefront curvature from Eq. (4.2). Exercises 4.6–4.8 give you some practice in applying this rule.

Moving forward, we make a change of variables in Eq. (4.16) to reveal that this integral is really just a 2D Fourier transform operation. Computer algorithms for performing 2D Fourier transforms are mature and fast, making this an attractive way of propagating paraxial beams with arbitrary amplitude and phase profiles through any set of optics. It is certainly much faster than evaluating the Huygens' integral 1.10 directly.

The changes of variables needed to rewrite Eq. (4.16) as a standard Fourier transform are

$$\begin{aligned} x' &= X' \sqrt{B\lambda} \\ y' &= Y' \sqrt{B\lambda} \\ x &= X\sqrt{B\lambda} \\ y &= Y\sqrt{B\lambda}. \end{aligned} \qquad (4.18)$$

With these substitutions Eq. (4.16) becomes

$$u(x, y, z) = ie^{-\frac{i\pi D}{B\lambda}(x^2 + y^2)} \iint_{\mathbb{R}^2} \left[e^{-i\pi A(X'^2 + Y'^2)} u\left(X'\sqrt{B\lambda},\, Y'\sqrt{B\lambda},\, z_1\right) \right] e^{i2\pi XX'} e^{i2\pi YY'} \, dX' dY' \qquad (4.19)$$

$$= ie^{-\frac{i\pi D}{B\lambda}(x^2 + y^2)} \, \mathbf{FT}\left[e^{-i\pi A(X'^2 + Y'^2)} u\left(X'\sqrt{B\lambda},\, Y'\sqrt{B\lambda},\, z_1\right) \right], \qquad (4.20)$$

where $FT[]$ indicates the 2D Fourier transform w.r.t. X' and Y'. (The result of the 2D Fourier transform is assumed to be oriented such that zero spatial frequency is at the center of the domain.)

Example 4.1 Fourier Propagation of a Gaussian Beam We now come a full circle and apply our 2D Fourier transform method to a Gaussian beam (TEM$_{00}$) mode to make sure it generates the expected downstream beam shape, presented without proof in Eq. (4.3). The complex field amplitude, $u(x, y, z)$, should be another Gaussian beam after propagating, with only a different amplitude, phase, and complex beam parameter q. We take the initial beam to have a waist with radius w_0 in the source plane at $z = 0$.

$$u(x, y, 0) = u_0 e^{-\frac{(x^2+y^2)}{w_0^2}} . \tag{4.21}$$

The initial beam parameter at $z = 0$ is $q_1 = \frac{i\pi w^2}{\lambda}$. From Eq. (4.4), we expect the beam parameter after propagation to be $q = q_1 + L$. We shall now see whether the Fourier transform method leads to the same result. Making the substitutions given in Eq. 4.18, gives

$$u(x, y, z) = i e^{-\frac{i\pi}{L\lambda}(x^2+y^2)} \iint_{\mathbb{R}^2} e^{-i\pi(X'^2+Y'^2)} e^{-\frac{L\lambda}{w_0^2}(X'^2+Y'^2)} e^{i2\pi XX'} e^{i2\pi YY'} dX' dY'. \tag{4.22}$$

I've performed the integral analytically with the help of the Mathematica computer program. I get

$$u(x, y, z) = \frac{1}{1 - \frac{iL\lambda}{\pi w_0^2}} e^{-i\pi(X^2+Y^2)} e^{i\frac{\pi^2 w_0^2}{\pi w_0^2 - iL\lambda}(X^2+Y^2)}, \tag{4.23}$$

which doesn't look too promising at first. However, some algebraic wrangling (see Exercise 4.3) confirms that this is indeed the same as Eq. (4.3)

$$u(x, y, z) = \frac{q_1}{q_1+L} e^{-ik\frac{x^2+y^2}{2(q_1+L)}} . \tag{4.24}$$

So, for Gaussian beams, the Fourier transform method, which is really just Huygens' integral in the Fresnel approximation, does agree with the simpler propagation method using the beam parameter q.

Usually, the Fourier transform is performed numerically rather than analytically. An example is shown in Figure 4.4, which was produced using the numerical Fourier propagation code in Appendix B.4. Once the code is set up, it's not difficult to change the source field. So, using the same code we can investigate the effect of any source-plane field amplitude we wish. Also, since the code returns the field amplitude in another plane, the field plane, this can become the source plane for subsequent iteration of the code. That way we can keep propagating the field in steps using any distance increment that's convenient. For example, if one wants to see what is the effect on the beam of mild flatness defects in a mirror, this can be investigated. First propagate the field to the mirror, then multiply the field amplitude by an (x, y)-dependent phase shift representing the flatness deviations. Then propagate through the rest of the optics chain. In this manner, it is possible to build up a

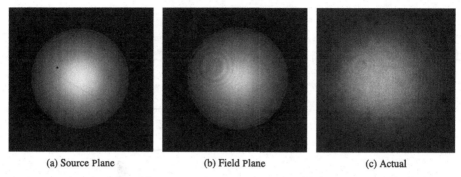

(a) Source Plane   (b) Field Plane   (c) Actual

Figure 4.4 Output of the Fourier transform code in Appendix B.4. The electric field in the field plane is related to the electric field in the source plane by the paraxial Huygen integral. The integral has been evaluated with a 2D Fast Fourier Transform according to Eq. (4.20). The source plane irradiance, shown in panel **(a)**, contains a beam with Gaussian profile obscured by a small (50 μm) piece of dust to the upper left of the beam axis. The field plane was chosen to be 5 cm downstream of the source plane and a divergent ($f = -3$ cm) lens was placed immediately after the source plane. The field plane irradiance is shown in panel **(b)**. Anyone who has viewed a beam that has passed through an optic with a speck of dust on it will recognize the characteristic diffraction pattern! Panel **(c)** shows an actual photo of this problem that was subsequently fixed with a puff of canned air.

realistic model of complex optics chains including realistic optical imperfections, aperture clipping, and so forth – a rather powerful tool.

4.2 Gratings and Spectrometers

Spectroscopy is a very important practical application of diffraction. It's no surprise then that the manufacture of gratings forming the heart of the spectrometers, is highly evolved. There are two basic types of grating: Transmission gratings and reflection gratings. Transmission gratings are made by scratching, etching, or depositing, closely spaced straight lines onto a transparent substrate (usually an optical quality glass). The lines block the light and it can only pass through the grating in the areas between the lines. Reflection gratings can be made similarly but in reverse, for example, by etching straight lines into a reflective coating deposited on a substrate. These two types of grating really behave the same way, except that one is reflective and the other is transmissive. Figure 4.5 shows a typical transmission grating. The angles at which such a grating diffracts light are those for which the wavefronts from adjacent gaps are in phase. The construction in Figure 4.5 indicates that (for beams entering the grating at normal incidence) the diffraction maxima occur whenever

$$\sin \theta = m\frac{\lambda}{a}, \qquad m = 0, 1, 2, \dots \qquad (4.25)$$

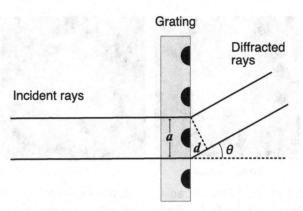

Figure 4.5 A small portion of a transmissive diffraction grating. The dark hemispheres represent the grating ruling. We get constructive interference at angles θ where the pathlength difference d is an integer number of wavelengths.

where λ is the incident wavelength and a is the line spacing. Equation (4.25) is known as the grating equation. The number m is known as the diffracted order where the zeroth order corresponds to light that passes straight through the grating without deviation. Note also there are a finite number of diffracted orders for any diffraction grating. When $\left|\frac{m\lambda}{a}\right| > 1$ there is no longer a solution to the grating equation. The more closely spaced the lines in the grating, the fewer diffraction orders will be available but the range of incident wavelengths will be spread over a correspondingly larger range of angles.

There is a third rather important type of grating that is known as a blazed grating. In this type of grating, rather than transmissive or reflective lines, the grating surface is made up of a series of reflective, uniformly spaced, parallel, saw-tooth shaped steps that extend the full height of the grating. The angle of the steps with respect to the grating surface, known as the blaze angle, determines the direction into which most of the light is reflected. Since the diffraction angle is still given by the grating equation above, the choice of blaze angle allows grating designers to cause a particular order m or range of orders to receive most of the diffracted light. Other wavelengths and orders still diffract as normal but are much dimmer. Using the incident light efficiently in this manner is especially useful if the spectrometer is starved for light, such as in astronomical observations. In astronomy, the gratings may also be very large in order to get sufficient light throughput.

Exercises

4.1 Show that for a Gaussian beam ($m = 0$, $n = 0$) passing through a waist with radius w_0 at $z = 0$ Eq. (4.1) is simply

$$u(x, y, 0) = u_0 e^{-\frac{x^2 + y^2}{w_0^2}}.$$

Hint: $R \to \infty$ at the waist.

4.2   Verify that the substituting Eq. (4.18) into Eq. (4.16) leads to Eq. (4.19).

4.3   Do example 4.1 for yourself, including the "algebraic wrangling." Thereby convince yourself that the Gaussian beam stays a Gaussian beam upon propagation. (It's worth knowing, if you didn't already, that the Fourier transform of a Gaussian is another Gaussian. Here you verify that in two dimensions.)

4.4   Modify Eq. (4.25) to cover the case where light is incident on the grating at an angle ϕ with respect to the grating normal. *Hint*: The incident rays in Figure 4.5 now travel different distances to reach the grating. This distance difference has to be taken into account in addition to d.

4.5   By integrating the irradiance from Eq. (4.14) over the xy-plane at some arbitrary location z on the optic axis, find the total power in a Gaussian beam (TEM_{00}) traveling in the $+\hat{z}$ direction. Now, imagine moving a vertical (parallel to $\hat{y}$) knife-edge across the beam from negative to positive x. Find the relative power transmitted past the knife-edge as a function of the edge position x. Plot the result. How could you use this device to find the radius w of a beam?

4.6   A Gaussian beam from a red HeNe laser comes to a waist immediately before entering a plano-convex lens of focal length $f = 1$ m. Since the phasefronts at the waist position are flat, the radius of curvature of the beam immediately before entering the lens is infinite. Therefore, immediately before entering the lens, the beam parameter q is purely imaginary. Take the value of q immediately before entering the lens to be $q = 496.3i$. What is the beam radius w_0 at the waist immediately before the lens? The lens will bring the beam to a new waist at a distance d from the lens. Find d and the beam radius w_0' at the new waist. Does the beam come to a focus (waist) at the nominal focal length?

4.7   A collimated Gaussian beam is focused by a positive lens. (A collimated Gaussian beam is going through a waist and has planar wavefronts.) Show that the beam radius at the focus (the new waist) is inversely proportional to the focal length of the lens.

4.8   A collimated Gaussian beam from a HeNe laser with beam radius w enters a Galilean telescope constructed of two lenses. The first lens is negative with focal length $-f_1$. The second lens is positive with focal length f_2. The distance between the lenses is $L = f_2 - f_1$. Find the beam radius w' immediately after the telescope. Is the beam collimated when it leaves the telescope? (A collimated Gaussian beam is going through a waist and has planar wavefronts.)

4.9   (Computer problem) In the xy–plane, plot wavefronts of the TEM_{00} and TEM_{05} modes as they travel through a focus. Plot both modes on a single graph using different colors or line styles to differentiate the modes. Choose your axes so that the optic axis is the z–axis and the focus is at $z = 0$. The domain of your plot should be $-2z_0 < z < 2z_0$, were z_0 is the Rayleigh range and $-w(2z_0) < y < w(2z_0)$, where $w(z)$ is the beam radius at z. *Hint*: Consider the argument of Eq. (4.13). Wavefronts can be defined as the locations where the phase is any integer multiple of 2π.

4.10 (Computer problem) Consider a Gaussian beam (TEM$_{00}$) traveling in the $+\hat{z}$ direction where the z–axis is the optic axis and the beam waist is at $z = 0$. At the origin, the irradiance is $I(0, 0, 0) = 1 \frac{W}{m^2}$. The wavelength is $\lambda = 632.8$ nm and the waist size is $w_0 = 3.0$ μm. Make a contour plot of the irradiance in the xz–plane, showing the following contours of the irradiance (in $\frac{W}{mm^2}$): $0.9, 0.3, 10^{-1}, 10^{-2}, 10^{-3}, 10^{-4}, 10^{-5}$. Explain why the contours are closed loops centered on the origin.

4.11 (Computer problem) Use the code in Appendix B.4 to calculate the diffraction pattern from a small circular aperture. Then do the same for a circular obstruction. Can you spot the spot of Arago?

4.12 (Computer problem) Find the diffraction pattern of vertically polarized FM waves (100 MHz) over the edge of the roof of a steel building lying directly in the path of the broadcast signal. You can model the building as a horizontal "knife-edge" obstruction. Can this explain any reception behavior you're familiar with from driving?

4.3 Experiment: Diffraction and Spectroscopy

Objectives

1 Recognize common diffraction patterns.

2 Compare theory to the observed diffraction from a circular aperture.

3 By calibrating an analog spectrometer, investigate the grating equation in a real-world instrument.

4 Estimate the energy in the spin-orbit coupling associated with the sodium D doublet.

5 Identify the spectral signature of Rayleigh scattering in the blue sky.

Equipment Needed

- Low-power visible laser such as a 1–5 mW HeNe.
- Pinholes with sizes ranging from about 10 to 100 microns (can be constructed with foil).
- Slits with widths ranging from about 10 to 1,000 microns (can be constructed with foil).
- Knife-edge
- A very small photodiode or a larger photodiode behind a small (<1 mm^2) aperture. A photodiode amplifier and readout are also needed. Alternatively, a good quality CCD/CMOS camera can be used.
- Good quality analog spectrometer with a grating of 1,200 lines/mm or similar.
- Fiber-coupled digital spectrometer covering most or all of the visible range.

> *Looking directly at the sun is harmful to your eyes. Be especially careful with any lenses you may be using. **Under no circumstance should you look at the sun through a lens or any optical device not explicitly designed for the purpose.** Doing so is very likely to lead to vision loss.*

Fresnel versus Fraunhofer Diffraction

Illuminate a variety of apertures (slits, pinholes, and knife-edges) with a laser and investigate two diffraction regimes depending on the width of the slit: Fraunhofer diffraction and Fresnel diffraction. Fraunhofer diffraction is the far-field limit of Fresnel diffraction. You can tell the difference between Fresnel and Fraunhofer diffraction by the fact that in Fresnel diffraction, the qualitative structure of the diffraction pattern changes as the aperture size d and/or distance to the screen D are changed. In Fraunhofer diffraction, the qualitative features of the pattern do not depend on d and D. Only the size of the Fraunhofer diffraction pattern changes.

> *Recommendations*: If premade apertures, slits, knife-edges, and so forth, are not available, you can construct your own. Razor blades make excellent knife-edges and are also good for making narrow slits in aluminum foil. When done carefully, the width of the slit can be adjusted by choosing a surface of appropriate hardness on which you perform the cut. Pinholes may be made by piercing aluminum foil with a needle or similar

sharp objects. Again, the diameter of the hole can be varied by choosing the hardness of the underlying surface. It can be helpful to have a short focal length *negative* lens on hand to expand the diffraction pattern near an aperture in order to make it easier to view on a screen or camera image sensor.

Airy's Disk

Find or construct a small circular aperture with diameter about 30–100 microns. Measure the irradiance of the diffraction pattern on a distant screen as a function of radius. This can be done with a small photodiode traversed over the pattern, or with a camera and appropriate image processing. If you choose to use a photodiode, make sure the image is large enough so that the photodiode is small compared to the width of any features. (If a very small photodiode is not available, a larger one can be masked leaving only a small clear aperture.) Make sure the room is dark during the measurements *and* subtract the background light level from each measurement. If you choose to use a camera, you must be able to control the beam power and/or the camera aperture so as to prevent image sensor saturation. Since the range of irradiance you are trying to measure is probably larger than the dynamic range of the image sensor, you may need to intentionally cause the image sensor to saturate on high irradiance areas of the diffraction pattern when you make measurements of low-irradiance areas. (This works with CMOS cameras but doesn't work with CCD cameras.) Then reduce the aperture of the camera or reduce the light level to measure the high irradiance areas. You can stitch the images together with an appropriate multiplication factor applied to match the levels at the appropriate locations. This diffraction pattern is known as the Airy pattern. Measure the distance from the aperture to the screen and make a note of the nominal size of the aperture, if available. The irradiance should be

$$I(r) = \left(\frac{2J_1(kar/L)}{kar/L}\right)^2 I_0, \qquad (4.26)$$

where I_0 is the maximum irradiance that occurs at the center of the diffraction pattern. J_1 is a Bessel function of the first kind, $k = 2\pi/\lambda$ is the wave number, a is the aperture radius, r is the radial distance from the center of the diffraction pattern and L is the distance from the aperture to the screen. By fitting your data to this function using a and I_0 as fit parameters, find the aperture radius a and the uncertainty in the aperture radius a. What deviations do you see from the expected behavior?

Recommendations: To get good resolution in your measurement of the diffraction pattern, you may have to project the beam over a significant distance. Do not use a divergent lens to expand the pattern or the formula mentioned earlier will not give you the correct aperture radius. During set up, do a quick check of the location of the first minimum in the Airy pattern to make sure everything is set up correctly. It should occur at an angle $\theta \approx 1.22\frac{\lambda}{a}$.

Analog Spectrometer

In this section you will familiarize yourself with the use and calibration of an analog spectrometer. In a later section of the lab you will use a digital spectrometer. In most work, digital spectrometers have supplanted analog ones. Like modern telescopes, digital spectrometers use image sensors to monitor the light passed through the device. (In the case of a laboratory spectrometer, the CCD/CMOS chip may be longer than it is wide in order to accommodate the diffraction pattern from the grating.) The advantage of using a digital spectrometer readout is that it allows one to easily store the diffraction pattern, also known as "the spectrum" of the light passing through the spectrometer. The disadvantage is that intuitive contact with the instrument is lost, especially since most commercial versions are sealed and don't allow the user to see how they work. Therefore, in this laboratory, you will start with an analog spectrometer and then graduate to the digital spectrometer. Use of the analog spectrometer also makes it much easier to recognize instrument systematics that may also be present in the digital version. Figure 4.6 shows a typical analog spectrometer with the grating placed on the central platform (cover removed).

In theory, the grating equation can be used to convert the measured diffraction angle to wavelength. Unfortunately, translations and rotations of the grating from the ideal position introduce "corrections" to the grating equation. Also, transmission gratings are etched on glass of nonzero thickness, which also affects the results slightly. Therefore, it's necessary to calibrate the spectrometer with a source of known wavelengths. The results are then interpolated or used to fit for the unknowns in a modified grating equation.

Spectrometer Calibration

Set up an analog spectrometer. You may need to see the instructor about installing the grating correctly. Obtain a narrow but workable slit width and figure out how to focus the slit and the reticle, if present. Make sure both are in focus. Once you have the spectrometer working, calibrate it using discharge tubes as a light source.

> *Be very careful not to touch the terminals of the discharge tube apparatus. **The terminals are at high voltage and may be capable of delivering a lethal electric shock.** Be especially careful with older models and noncommercial models that may not be properly current limited.*

Each discharge tube contains a low-pressure vapor of a single element, such as Hydrogen, Helium, Sodium, Krypton, Mercury, and so forth. When a current is forced through this low-pressure vapor, valence electrons are excited to higher states. As the electrons return to their equilibrium states in a series of jumps between energy levels, they emit characteristic wavelengths. These wavelengths can be used to calibrate your spectrometer. Tables showing the emission wavelengths of common elements are maintained by the US National Institute of Standards and Technology (NIST) and are available online via their "Atomic Spectra Database."

Figure 4.6 An analog spectrometer seen from above. Light enters through a small vertical slit
capping the entrance to the tube at the top of the picture. (An adjustment screw
controlling the slit width can be seen protruding from the black cap holding the slit.)
Light passes through the tube and is incident on the grating placed on the small
platform at the center of the instrument. The grating diffracts the light off to the lower
left at an angle that depends on wavelength. The tube at the lower left contains an
eyepiece that captures the diffracted light and (in conjunction with the eye) images the
entrance slit onto the retina. The eyepiece may also contain a reticle or cross-hairs for
accurate alignment and reading of the spectrum. To capture different wavelengths, the
tube with the eyepiece is rotated about the center-post of the instrument. The plate at
the bottom rotates with the tube and may provide a vernier scale for accurate reading
of the angle/wavelength.

Recommendations: To calibrate, you will need to make a plot of the angle of the
spectrometer's output telescope versus emission line wavelength. Make sure you include error estimates for your angle measurements. You should then either interpolate
this data or fit it to a modified grating equation in order to convert angle values to
actual wavelength values. Interpolation requires more calibration wavelengths while

obtaining a modified version of the grating equation requires theoretical work but less data. You can try either approach, or both.

Doublets in the Sodium Spectrum

Measure the primary emission wavelengths of sodium using a sodium lamp. The lamp is delicate, so treat it gently. The lamp also gets very hot; under no circumstances should it be covered. In addition to the primary orange-yellow "Sodium D" emission lines, there are numerous others. Also measure the wavelengths of these. The Sodium D lines are due to electrons in the 3p orbital state decaying to the 3s state, each releasing a photon in the process. The presence of two lines rather than one is due to transitions of electrons in two different electron spin states corresponding to spin angular momentum quantum numbers of 1/2 and 3/2 respectively. Due to an interaction (a "coupling") between the electron spin and the orbital angular momentum of the electrons, the two spin states actually have slightly different orbital energies. Hence, there are two different wavelengths released corresponding to the two possible spin states of the electron. In this manner, spectroscopy can give us quantitative information about the interior of atoms. By measuring the distance between the lines in the doublet, measure the energy stored by the spin-orbit coupling. Quote the coupling energy in eV with uncertainties. Does it matter which doublet is used to make the estimate (i.e., sodium D doublet or some other doublet)? Why or why not? Also, compare the spin-orbit coupling energy in each case to the energy of the electron transition itself. Does this give you an estimate of the relative strength of the corresponding forces on the electron? (Do such questions even make sense?)

Rayleigh Scattering

Looking directly at the sun is harmful to your eyes. Be especially careful with any lenses you may be using. **Under no circumstance should you look at the sun through a lens or any optical device not explicitly designed for the purpose.** *Doing so is very likely to lead to vision loss.*

The sky is blue due to Rayleigh scattering, which is more effective at shorter wavelengths. Conversely, the spectrum of direct sunlight or the light passed through clouds is fairly constant in the visible band, peaking gently at about 500 nm. By taking the ratio of a blue-sky spectrum and a direct sunlight spectrum or a cloud spectrum, find the approximate dependence of Rayleigh scattering efficiency on wavelength. Compare it to the expectation from theory. (See, for example, Griffiths (2017).)

Recommendations: First, calibrate the digital spectrometer using discharge tubes in a manner similar to the calibration you made of the analog spectrometer. In this case though you are only correcting for any errors in the manufacturers' default calibration. You'll make a table of nominal wavelengths (as indicated by the spectrometer) versus actual wavelengths (as dictated by the element in the discharge tube). Then interpolate this table or fit to a low order polynomial to create a calibration conversion.

Getting enough scattered blue-sky light into the digital spectrometer should not be hard. Just pointing the end of the optical fiber at the sky may suffice. If not, you can always use a lens to collect light into the spectrometer. However, be careful not to focus direct sunlight into the spectrometer as this is likely to ruin it. A similar method works to obtain a spectrum of a cloudy sky.

Before attempting to obtain a spectrum of direct sunlight, make sure you remove any collecting lenses. Pointing the end of the fiber toward the sun should suffice. You may also have to place a neutral density filter over the end of the fiber so as not to saturate the spectrometer. (Another method to prevent saturation is to tilt the fiber slightly away while making sure your spectrum is still dominated by direct sunlight and not scattered light).

Ideas for Further Investigation

Compare the diffraction at a knife-edge to theory by imaging the diffraction pattern and fit a cut to an appropriate analytical model.

Compare the diffraction patterns from a variety of apertures to the results of numerical simulations employing the 2D Fourier transform method.

5 Interferometry

5.1 Interferometer Types

Interferometry refers to the practice of combining two or more beams in order to observe the interference pattern formed. The interference pattern reveals the differences between the wavefronts of the beams involved and can be used to monitor the change in wavefronts of one beam with respect to another. Interferometry is useful because it converts wavefront *phase* variations to irradiance variations on a screen or at a detector. Optical interferometers can detect variations in wavefront flatness or position at the picometer (10^{-12} m) level or better (sometimes much better). Compared with the basic resolution of a micrometer (10^{-6} m), an optical microscope (10^{-7} m), or even an electron microscope (10^{-10} m), the interferometer is a very sensitive device. However, it is fundamentally not an imaging device.

In interferometry, the beams that are combined to form interference originate from a single "input beam" that is split into two or more beams that form the "arms" of the interferometer. The beams forming the arms of the interferometer are later combined to produce the interference. It's common to classify interferometers according to the way in which the input beam was split. "Amplitude splitting" interferometers use partially reflective/partially transmissive optics like a non-polarizing beamsplitter (traditionally a mirror with a silver coating that is too thin to be fully reflective). The Michelson interferometer is a classic example of an amplitude splitting interferometer. "Wavefront splitting" interferometers separate different transverse regions of a single beam. The separated parts then travel along different paths before eventually recombining. A classic version of wavefront splitting is Young's double slit experiment where a beam is incident on a pair of closely spaced slits. The two beam parts emerging from the slits then recombine downstream. Another way to wavefront-split a beam is to pass the beam through a prism in such a way that an edge of the prism splits the output beam in two.

Interferometers can also be classified according to whether they interfere two beams or multiple beams. I prefer this approach because interferometers that split the input beam many times, even dozens of times, behave quite differently from ones that split the input beam just once.

5.1.1 Two-Beam Interferometers

The Michelson interferometer is a classic two-beam interferometer named after Albert A. Michelson, who in 1887 with Edward Morley, used it to rule out the existence of "aether"

– a putative medium in which light traveled. The Michelson interferometer works by splitting a single beam of coherent light (usually a laser these days) into two beams that then travel down two paths ("arms") of similar length before being retroreflected back to the beamsplitter where they are recombined (see Figure 5.1a). The recombined beams exit the beamsplitter as two beams, each of which contains light from both arms. The relative power in these beams depends on the relative phase of the beams returning from the arms. In the extreme cases where the relative phase between the beams returning from the arms is 0° or 180°, all the power is directed into one of the recombined beams while the other recombined beam is dark. We may choose to monitor only one of the recombined beams, or we can use both of the output beams for a balanced signal.

Following are some examples of the many other two-beam interferometers resembling the Michelson interferometer in layout and function.

- The Mach-Zehnder interferometer recombines the beams at a second beamsplitter rather than reusing the same beamsplitter twice.

- A Rayleigh refractometer is a wavefront-splitting version of a Jamin interferometer. Both are used to measure the phase difference accumulated by closely spaced parallel beams.

- The Fizeau interferometer is used to compare the surface figures of reflective surfaces.

- The Twyman-Green interferometer is similar to the Fizeau interferometer except it is sensitive to the phase accumulated in transmission through a lens.

A rather different class of two-beam interferometers exists where both beams take exactly the same path but traverse the path in opposite directions. Such interferometers are called Sagnac interferometers. If a Sagnac interferometer is rotated in the plane of the beams, the counter-propagating beams pick up differential phase and the interference pattern changes. This is due to the fact that the beamsplitter is approaching one of the beams but receding from the other; the counter-propagating beam's path is effectively shortened and the co-rotating beam's path is lengthened. This is known as the Sagnac effect. Ring laser gyroscopes rely on this effect and are an important practical application of the Sagnac interferometer.

Example 5.1 Phase Sensitivity of a Michelson Interferometer   Traditionally, the Michelson interferometer is intentionally misaligned slightly. The idea is that when the output beam is shone onto a screen, it should have several fringes. Fringes are the alternating regions of constructive interference (bright) and destructive interference (dark) shown in the figure below. As the armlength difference changes, these fringes move across the screen allowing them to be counted as they pass by. Nowadays we don't usually count fringes manually but rather place a photodiode at the output, especially if we are trying to measure very small armlength changes. We try to align the interferometer well enough that a single large fringe occupies the entire output beam. That way, when the armlength difference changes, the overall power of the output beam

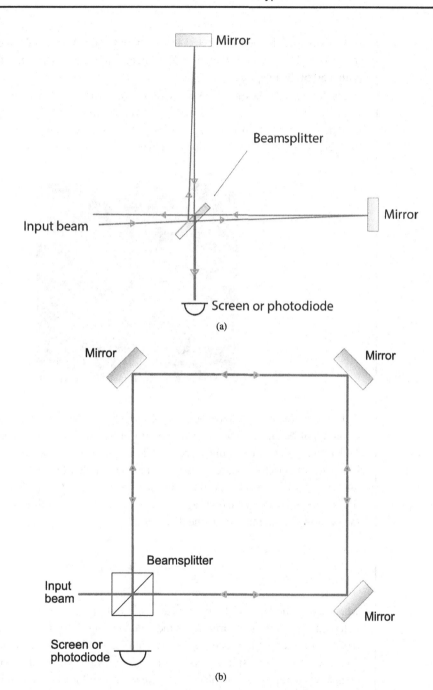

Mirror

Beamsplitter

Input beam

Mirror

Screen or photodiode

(a)

Mirror   Mirror

Beamsplitter

Input
beam

Mirror

Screen or
photodiode

(b)

Figure 5.1 Two-beam interferometers. **(a) Michelson interferometer**. One of the recombined beams exits to the left and is often not used. The other recombined beam is the "output beam" and is incident on the detector at the bottom of the figure. **(b) Sagnac interferometer**. The two beams from the beamsplitter rotate in opposite directions around the interferometer and recombine at the photodiode. As in the case of the Michelson, the other recombined beam travels back toward the laser and may or may not be used.

changes, which is then registered by the photodiode. We'll now calculate the rate of irradiance change as a function of armlength difference for a well aligned Michelson with a 50/50 beamsplitter.

The figure here shows fringes at the output of a Michelson interferometer. In addition to the clear vertical fringe structure, you can see that the bright fringes contain laser speckle. On the leftmost fringe there are also some additional fringes due to parasitic interference from a retroreflection somewhere in the optics chain.

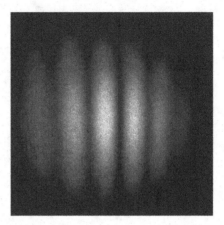

The output beam is a superposition of the two arm beams, each with half the power of the input beam. We will refer to the arm in line with the input beam as the x−arm. This is the horizontal arm in Figure 5.1a. The other arm will be the y−arm. Since power is proportional to the square of the field, the magnitude of the field in the arms is $\frac{1}{\sqrt{2}}$ of the input field magnitude. At the point of recombination (at the detector or screen) the fields from each arm have once again been split at the beamsplitter and so gain a second factor of $\frac{1}{\sqrt{2}}$ in magnitude. At the detector then

$$E_x = \frac{E_{in}}{2} e^{i(\omega t)} \qquad E_y = \mp \frac{E_{in}}{2} e^{i(\omega t - \phi)}, \tag{5.1}$$

where E_{in} is the field of the input beam before it is split by the beamsplitter, E_y is the field coming from the y−arm, E_x is the field coming from the x−arm, ω is the angular frequency, and t is time. The $\mp$ sign in front of E_y is due to the fact that reflection from one side of the beamsplitter leads to a negative sign while reflection from the other does not. Therefore, if one places the detector at the location shown in Figure 5.1a the negative sign applies. This is known as the antisymmetric port of the interferometer. If one places the detector at the other output, on the same side of the beamsplitter as the input beam, known as the symmetric port of the interferometer, the positive sign applies. The angle ϕ is the additional phase accumulated in the y−arm with respect to the x−arm due to their length difference. So ϕ will be positive if the y−arm is longer than the x−arm.

The sum of these fields at the detector, $E_{\text{det}} = E_x + E_y$, is multiplied by its complex conjugate to get a quantity proportional to the irradiance seen by the detector

$$I_{\text{det}} \propto E_{\text{det}}^* E_{\text{det}} \tag{5.2}$$

$$= \frac{E_{\text{in}}^* E_{\text{in}}}{4} \left(2 \mp e^{i\phi} \mp e^{-i\phi} \right) \tag{5.3}$$

$$= \frac{E_{\text{in}}^* E_{\text{in}}}{2} \left(1 \mp \cos \phi \right) \tag{5.4}$$

So

$$\frac{I_{\text{det}}}{I_{\text{in}}} = \frac{1 \mp \cos \phi}{2}. \tag{5.5}$$

The signal from a Michelson repeats each time ϕ passes through 2π, which we refer to as "fringe wrapping."

One normally operates a simple Michelson interferometer around the "center of the fringe." This means we try to keep ϕ as close to $\pm\pi/2$ as possible, or odd multiples thereof. In practice we achieve this by implementing electronic feedback to move one of the arm mirrors in order to keep the interferometer "on fringe." If the mirrors move so as to deviate from the center of the fringe, we'll call the corresponding phase deviation θ.

$$\phi = \pm\frac{\pi}{2} + \theta. \tag{5.6}$$

In terms of θ, the normalized output irradiance at the *antisymmetric* port is then

$$\frac{I_{\text{det}}}{I_{\text{in}}} = \frac{1}{2} \left[1 - \cos\left(\pm\frac{\pi}{2} + \theta \right) \right] \tag{5.7}$$

$$= \frac{1}{2} \left(1 \pm \sin \theta \right). \tag{5.8}$$

If the differential arm length deviation is ϵ, this leads to a phase deviation $\theta = \frac{4\pi}{\lambda}\epsilon$. The corresponding normalized output irradiance is then

$$\frac{I_{\text{det}}}{I_{\text{in}}} = \frac{1}{2} \left[1 \pm \sin\left(\frac{4\pi}{\lambda}\epsilon \right) \right]. \tag{5.9}$$

The rate of irradiance change per differential arm length change at the center of the fringe is then

$$\left. \frac{dI_{\text{det}}}{d\epsilon} \right|_{\epsilon=0} = \pm\left(\frac{2\pi}{\lambda} \right) I_{\text{in}}. \tag{5.10}$$

Sometimes it's more convenient to refer this to the irradiance $I_{\text{one\_arm}} = \frac{I_{\text{in}}}{2}$ incident on the detector when the light from the other arm is blocked. In that case, the slope is

$$\left. \frac{dI_{\text{det}}}{d\epsilon} \right|_{\epsilon=0} = \pm\left(\frac{4\pi}{\lambda} \right) I_{\text{one\_arm}}. \tag{5.11}$$

This formulation makes it really easy to calibrate your signal. Since the voltage out of your photodiode amplifier is proportional to the irradiance on the detector, you can simply block the light in one arm with a card and read the voltage $V_{\text{one\_arm}}$. Then the on-fringe voltage change per unit armlength change is

$$\left.\frac{\mathrm{d}V_{\text{det}}}{\mathrm{d}\epsilon}\right|_{\epsilon=0} = \pm\left(\frac{4\pi}{\lambda}\right)V_{\text{one\_arm}}. \tag{5.12}$$

Example 5.2 Fringe Contrast If the two beams that combine to produce the output don't have the same irradiance, they can't cancel perfectly. In that case, the destructive interference fringes will still have light in them, they will simply have less irradiance than the constructive interference fringes. The fringes are said to have a "contrast defect." The "fringe contrast" can be quantified as

$$\nu = \frac{I_{\text{max}} - I_{\text{min}}}{I_{\text{max}} + I_{\text{min}}} \tag{5.13}$$

and is sometimes also called the "fringe visibility." Here I_{max} is the maximum brightness of the bright fringes and I_{min} is the minimum brightness of the dark fringes.

Consider the case of any two-beam interferometer where the beam returning from one of the arms has 35% of the input power while the beam returning from the other arm has 55% of the input power. Here, we'll assume that the beams returning from the arms are of the same shape and size so their profiles overlap well. There are two things that occur here: some light is being lost in the arms and something is happening either at a beamsplitter or in the arms that causes a power imbalance. The power imbalance could be due to any number of things: non-50/50 beamsplitters, different mirror types in the two arms, absorption in any medium placed in the arms, polarization rotation in one arm with respect to the other, and so forth. The loss in the arms could be due to the same mechanism leading to the imbalance or it could be due to other mechanisms such as reflection of part of the input light on the front face of the beamsplitter. The interference pattern will have $I_{\text{max}} = (0.35 + 0.55)I_{\text{in}} = 0.9I_{\text{in}}$ and $I_{\text{min}} = (0.55 - 0.35)I_{\text{in}} = 0.2I_{\text{in}}$. The fringe contrast is then

$$\nu = \frac{0.9 - 0.2}{0.9 + 0.2} = 0.63.$$

A set of fringes with visibility of 63% would look fairly bad on a screen. One could obtain better fringes by putting a wedge in the arm with more light in order to reflect out some of the light from that arm and obtain better contrast.

5.1.2 Multiple-Beam Interferometers

Multiple-beam interferometers typically use partially reflective optical surfaces to split a beam sequentially into multiple beams that are then combined. A prototypical multiple-beam interferometer is the original Fabry-Perot etalon consisting of two parallel, partially reflective, flat mirrors illuminated by a slowly diverging source of monochromatic light (such as a filtered and collimated low pressure mercury discharge lamp or a laser). The setup is shown in Figure 5.2 along with the fringes produced. The fringes are circular (due

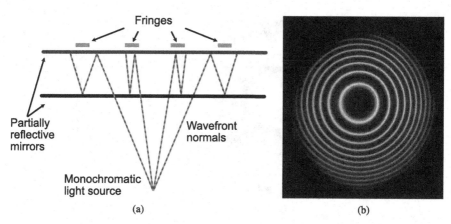

Figure 5.2 A Fabri-Perot etalon. **(a)** Side-view. The wavefront normals (rays) reflect within the etalon leading to multiple output rays. Bright fringes occur when rays enter at angles for which the cavity traversal induces an integer-wavelength path difference between consecutive transmitted rays. When viewed by eye or with a camera, the output consists of concentric circular fringes as shown in **(b)**. The fringes shown are due to a mercury vapor lamp filtered to emit only the green line at 546 nm.

to axial symmetry) and occur for rays traveling at angles where the pathlength between the plates is an integer number of half-wavelengths.

A resonant optical cavity, discussed in Chapter 7, is another example of a multiple-beam interferometer. In this case, the multiple beams are generated by amplitude splitting at the input mirror and are fully overlapping as they enter, resonate, and exit the cavity.

Perhaps the most important use of multiple-beam interferometry is in multilayer optical coatings. Multilayer coatings are useful in a large range of optics applications such as antireflective coatings, mirrors, optical filters, and so forth. High-performance mirrors intended for a particular wavelength of light are made by depositing multiple, quarter wavelength thick layers of transparent dielectric materials onto a substrate (usually a glass blank; see Figure 5.3). Every other layer has a high index of refraction while the remaining layers have a low index of refraction. Often, the low-index layers are made of silica glass while the high-index layers are made of an amorphous metal oxide. At each interface between layers there is a an index change. Therefore, part of any incident beam will be retroreflected at each layer interface. The full reflected beam is just the sum of all these partial reflections. Since the layers are all a quarter wavelength thick and there is a sign change on reflection at every other interface, all the partial reflections are *in phase* with one another. This constructive interference between all the partial reflections can lead to much higher reflectivities than are available from metal mirrors. The reflectivity of such mirrors at their center wavelength λ_0 can be exceedingly high. Mirrors with reflectivities as high as 99.999% are commercially available and even higher reflectivities have been demonstrated for research purposes.

The formulae for evaluating the theoretical reflectivity of a stack of dielectric layers are fairly complex and make use of so-called transfer matrixes where a four-element matrix

 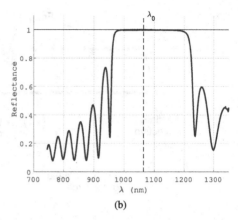

(a)   (b)

Figure 5.3 **(a)** A high-reflective (HR) multilayer coating, also known as a quarter-wave stack or Bragg reflector. This mirror is highly reflective at 1064 nm and mostly transmissive in-visible. The reflection we do see is due to one or more of the minor reflectance peaks away from the high-reflectance band. **(b)** Reflectance of an HR mirror as a function of wavelength. This particular mirror was designed to be highly reflective at a $\lambda_0 = 1064$ nm for use with a NdYAG laser. It has 30 layers (15 high-/low-index layer pairs). The high-index layer material is amorphous tantalum pentoxide and the low-index material is amorphous silica (silica glass).

represents the action of a single layer, and a corresponding matrix product represents the stack as a whole. For example, at normal incidence, the reflection coefficient of a stack of alternating high- and low-index dielectric layers is

$$r = \left(\frac{an_0 + bn_0n_s - c - dn_s}{an_0 + bn_sn_0 + c + dn_s} \right), \tag{5.14}$$

where n_0 is the index of the incident medium (usually air) and n_s is the index of the substrate (often glass). The coefficients a, b, c, and d are given by[1]

$$\begin{pmatrix} a & b \\ c & d \end{pmatrix} \equiv \left[\begin{pmatrix} \cos(\frac{2\pi}{\lambda}n_H\ell_H) & -\frac{i}{n_H}\sin(\frac{2\pi}{\lambda}n_H\ell_H) \\ -in_H\sin(\frac{2\pi}{\lambda}n_H\ell_H) & \cos(\frac{2\pi}{\lambda}n_H\ell_H) \end{pmatrix} \begin{pmatrix} \cos(\frac{2\pi}{\lambda}n_L\ell_L) & -\frac{i}{n_L}\sin(\frac{2\pi}{\lambda}n_L\ell_L) \\ -in_L\sin(\frac{2\pi}{\lambda}n_L\ell_L) & \cos(\frac{2\pi}{\lambda}n_L\ell_L) \end{pmatrix} \right]^N,$$

where n_H and n_L are the high and low layer indices, respectively; the physical layer thicknesses are ℓ_H and ℓ_L, respectively. We are assuming that the outermost layer is a low-index layer and there are a total of $2N$ layers, (N layer pairs). For a quarter-wave, high-reflective, stack we choose $n_H\ell_H = n_L\ell_L = \lambda_0/4$ where λ_0 is the desired "center wavelength" at which the coating is most reflective. The result is a curve similar to the measured reflectivity data

---

[1] See, for example, Fowles (1989).

shown in Figure 5.3b (see Exercise 5.8). For a quarter-wave stack of N layer pairs, the reflectivity at the center wavelength λ_0 of the coating can be shown to be

$$R = \left[\frac{\left(\frac{n_H}{n_L}\right)^{2N} - 1}{\left(\frac{n_H}{n_L}\right)^{2N} + 1} \right]^2. \tag{5.15}$$

5.2 Selected Applications of Interferometers

5.2.1 Fourier-Transform Spectroscopy

It turns out that a Michelson interferometer can be used as a high-resolution spectrometer by varying the length of one arm and recording the interferometer output. The output is known as the interferogram and its Fourier transform gives the spectrum of the light in the interferometer. This application was in fact discovered by Michelson himself who also built a mechanical device to take the Fourier transform of the interferogram![2] Fourier-transform spectrometers benefit from more light throughput than traditional, grating-based spectrometers, and so are inherently less noisy. They've found the greatest use at infrared wavelengths where they are known as Fourier Transform Infrared (FTIR) spectrometers.[3] Figure 5.4 shows the layout and operation of a basic Fourier-transform spectrometer.

If we move the x-arm mirror of a Michelson toward the beamsplitter by a distance, ϵ, the corresponding optical path length reduction is $x = 2\epsilon$ since the beam goes both up and down the arm. The induced differential arm phase is $\phi = \frac{2\pi x}{\lambda} = kx$ and Eq. (5.5) gives the corresponding interferometer response

$$\frac{I_{\text{det}}}{I_{\text{in}}} = \frac{1}{2} \left[1 - \cos\left(kx\right) \right]. \tag{5.16}$$

Now consider the case where the interferometer light consists of two different wavelengths λ_1 and λ_2 with corresponding wavenumbers k_1 and k_2 and angular frequencies ω_1, ω_2. Then the fields at the antisymmetric port are

$$E_x = \frac{E_1}{2} e^{i\omega_1 t} + \frac{E_2}{2} e^{i\omega_2 t}. \tag{5.17}$$

$$E_y = -\frac{E_1}{2} e^{i\omega_1 t - \phi_1} - \frac{E_2}{2} e^{i\omega_2 t - \phi_2}. \tag{5.18}$$

Multiplying the sum of these fields by their complex conjugate to get the total irradiance at the detector, as before, gives (after a long algebra slog)

$$I_{\text{det}} = \frac{I_1}{2} + \frac{I_2}{2} - \frac{I_1}{2} \cos\phi_1 - \frac{I_2}{2} \cos\phi_2 + \frac{E_1 E_2}{2} |z| \cos\left[(\omega_2 - \omega_1)t + \text{Arg}\,(z)\right], \tag{5.19}$$

[2] See Lecture IV in Michelson (1902).
[3] Several books cover Fourier-transform spectroscopy. This treatment follows the approach of Pedrotti (2007). A slightly different approach can be found in Bennett (2008) and a qualitative discussion of the strengths and limits of Fourier-transform spectroscopy can be found in James (2014).

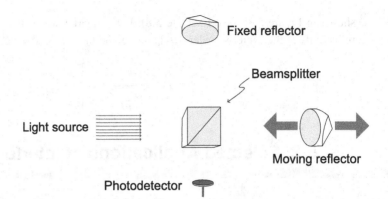

Light source

Fixed reflector

Beamsplitter

Moving reflector

Photodetector

Figure 5.4 A basic Fourier-transform spectrometer. The spectrum of the input light is measured as the length of one of the arms is changed by a large amount. Corner cubes (or trihedral prisms) may be used in lieu of a mirror because the beam always exits the cube in the same direction as it entered, regardless of the corner cube angular alignment. That reduces the engineering challenge associated with moving the reflector long distances without changing the interferometer alignment. The farther the reflector is moved, the higher the spectrometer resolution. The more often the reflector's position is read, the greater the spectrometer's bandwidth. The position of the reflector can be read out with an auxiliary laser passing through the same interferometer.

where $z = 1 + e^{i\phi_1} + e^{-i\phi_2} + e^{i(\phi_1 - \phi_2)}$ is a constant complex number. If we now take the time average of this quantity over many periods of the cosine, the cosine term will contribute zero. (In practice, one would be recording the output of the interferometer on a photodiode that can't respond at optical frequencies, so the cosine term with frequency $\omega_2 - \omega_1$ would not be registered anyway.) Using $I_{in} = I_1 + I_2$, we are left with

$$\frac{I_{det}}{I_{in}} = \frac{1}{2} - \frac{1}{2}\left[\frac{I_1}{I_{in}}\cos(k_1 x) + \frac{I_2}{I_{in}}\cos(k_2 x)\right]. \tag{5.20}$$

In other words, scanning the interferometer gives an interferogram that is the sum of two sinusoidal waves with wavenumbers corresponding to the wavelengths of the colors in the beam. If instead of just two wavelengths, we have a continuum of wavelengths with fractional irradiance contributions $f(k)dk = \frac{I(k)dk}{I_{in}}$, the sum becomes an integral

$$\frac{I_{det}}{I_{in}} = \frac{1}{2} - \frac{1}{2}\int_0^\infty f(k)\cos(kx)\,dk. \tag{5.21}$$

The function $f(k)$ is the spectral density of the light. It is the quantity we ultimately want to find. By definition, $f(k)$ is positive, real, and only defined for positive wavenumbers k. However, it'll be mathematically easier for us if we extend this to negative wavelengths by reflecting it about the origin to create an even function: $f(-k) \equiv f(k)$. In the end, we can always just ignore the $-k$ parts when interpreting $f(k)$ physically. So

$$\frac{I_{det}}{I_{in}} = \frac{1}{2} - \frac{1}{4}\int_{-\infty}^\infty f(k)\cos(kx)\,dk. \tag{5.22}$$

We can add zero to the second term without affecting the equation. We do so in the following way

$$\frac{I_{det}}{I_{in}} = \frac{1}{2} - \frac{1}{4} \int_{-\infty}^{\infty} f(k) \left[\cos(kx) + i\sin(kx)\right] dk \qquad (5.23)$$

$$= \frac{1}{2} - \frac{1}{4} \int_{-\infty}^{\infty} f(k) e^{ikx} dk. \qquad (5.24)$$

The second term in the first line is odd and so contributes nothing to the integral, which is over a symmetric domain. Apart from the factor of $\frac{1}{4}$, the second term in Eq. (5.24) is just the standard Fourier transform of $f(k)$. It can be inverted to give $f(k)$, which is ultimately what we wanted.

$$f(k) = \frac{2}{\pi} \int_{-\infty}^{\infty} -\left(\frac{I_{det}}{I_{in}} - \frac{1}{2}\right) e^{-ikx} dx. \qquad (5.25)$$

The subtraction of the $\frac{1}{2}$ serves to remove the average. In practice, we can be lazy and neglect to do this. The effect is a delta function "spike" on the spectrum at $k = 0$, which we just ignore. Also, we don't really need to worry about the sign of the integrand; since we know $f(k)$ is positive, we can just take the absolute value of the right-hand side. Finally, if all we really care about is the relative spectrum and I_{in} is constant, it leaves us with the wonderfully simple recipe

$$f(k) \propto \left| \int_{-\infty}^{\infty} I_{det} e^{-ikx} dx \right| \qquad (k > 0). \qquad (5.26)$$

The Fourier transform can be generated on a computer, in real time if need be, giving the spectrum directly.

The integral in Eq. (5.26) indicates that to get all the spectral components, we would have to know the interferogram $I_{det}(x)$ for $0 \le x < \infty$. In practice, the pathlength change of the arm, x, can't become infinite and this sets the limit on the resolution with which the spectrum can be measured. The wavelength resolution can be shown to be

$$\Delta\lambda = \frac{\lambda^2}{x_{max}}. \qquad (5.27)$$

For a modest 1 cm of mirror motion and near infrared wavelengths $\lambda \approx 1$ μm, we get $\Delta\lambda \approx$ 0.1 nm. The wavelengths accessible by this technique are also limited by the resolution with which the interferogram, I_{det} is measured. The more data points one acquires in the interferogram, the higher the accessible wavelengths. The number of data points N required to get spectral information down to a certain wavelength λ_{min} can be shown to be

$$\lambda_{min} = 2\frac{x_{max}}{N}. \qquad (5.28)$$

So, if we have 1 cm of mirror motion, we would at least need 10,000 data points to reach the near infrared. That means we have to be able to record the mirror position to within 1 μm while moving the mirror and maintaining the alignment to a high degree of accuracy. These requirements constitute a moderate instrument-building challenge.

Example 5.3 Interferogram of a Bandpass Filter Consider light filtered through a bandpass filter that has a Gaussian transmission profile. What is the interferogram produced with such a source? The light's spectral density is

$$f(k) = \frac{I_{\text{in}}}{\sqrt{\pi}\sigma} e^{-\frac{(k_0-k)^2}{\sigma^2}} \qquad (k \geq 0), \tag{5.29}$$

where k_0 is the center wavenumber and σ is the standard deviation of the transmission profile that has characteristic bandwidth $\Delta k = 2\sigma$. Although this Gaussian transmission profile is specified in terms of wavenumber k, it will also be approximately Gaussian in wavelength λ provided the width is small. Here's the corresponding transmission as a function of wavelength where I've chosen the center wavelength at 532 nm and full width of the transmission profile as 40 nm.

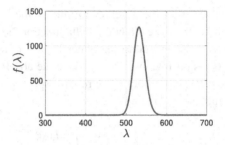

Now, we'd like to use Eq. (5.24) to find $\frac{I_{\text{det}}}{I_{\text{in}}}$ explicitly because we remember that the Fourier transform of a Gaussian is another Gaussian, so it seems likely that the integral is at least possible to do. In order to use this, however, we do need to extend $f(k)$ to negative $k's$ by reflecting it about the origin. The extended $f(k)$ is then

$$f(k) = \frac{I_{\text{in}}}{\sqrt{\pi}\sigma} \left[e^{-\frac{(k_0-k)^2}{\sigma^2}} + e^{-\frac{((-k_0)-k)^2}{\sigma^2}} \right], \tag{5.30}$$

which looks like this:

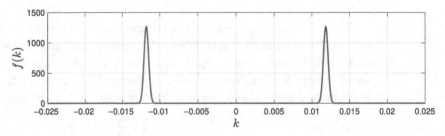

The integral in Eq. (5.24) can be done with the help of integral tables or computer programs implementing such tables (Mathematica, Maple, Wolfram Alpha, etc.). The result is

$$\frac{I_{\text{det}}}{I_{\text{in}}} = \frac{1}{2} - \frac{1}{2} \cos(k_0 x) e^{-\left(\frac{\sigma x}{2}\right)^2}. \tag{5.31}$$

So, this interferogram is a sine-Gaussian and looks like this:

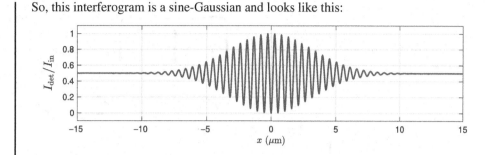

5.2.2 Gravitational Wave Astronomy

Gravitational waves are the gravitational analog of electromagnetic waves. The conceptually simplest gravitational wave emitter is a pair of masses joined by a spring and vibrating harmonically around the center of mass. Due to momentum conservation, there can be no dipole moment of the source and the gravitational radiation emitted is quadrupole radiation. The strain field has a quadrupolar pattern as shown in Figure 5.5. The polarization shown in the figure is the "+" polarization and the other possible polarization is the "×" polarization, identical except that it's rotated by 45° about the direction of travel.

A similar source of gravitational waves is two masses attached to the two ends of a rod and whirled about the center of mass. These will also radiate because circular motion is just the sum of linear harmonic motion in two orthogonal directions with a $\pi/2$ phase shift between the directions. Unfortunately, it's not hard to show that any conceivable man-made oscillator of this type would radiate far too weakly to detect with the current technology. However, using large Michelson interferometers, gravitational waves from extremely compact and massive objects in the universe have been detected: black holes and neutron stars whirling round one another in extremely close orbit, and subsequently merging.[4] Other sources, such as rotating single neutron stars and supernovae explosions, may also be detectable in the near future.

The interferometers used to detect such gravitational waves are the most sensitive length measurement devices ever constructed. Strong gravitational waves detected to date shake the interferometer mirrors with amplitudes on the order 10^{-18} m (approximately $\left(\frac{1}{800}\right)$th of a proton diameter). If the gravitational wave is well-oriented with respect to the interferometer arms then the mirror motion causes a differential arm length change, which is proportional to the gravitational wave strain. (This corresponds to a difference in the time-of-flight of photons in the two arms and a corresponding phase difference between the beams arriving at the beamsplitter.) So, the interference condition at the antisymmetric output of the interferometer changes and the amount of light passed to the output port

---

[4] See the seminal first-detection paper by the LIGO and Virgo collaborations: LIGO and Virgo Scientific Collaborations et al. (2016). I also recommend an amazing paper describing observations in gravitational and electromagnetic waves of the inspiral and merger of a binary neutron star system: 52 teams of gravitational wave, electromagnetic, and neutrino astronomers et al. (2017). For up-to-date information about current detections, see the LIGO Scientific Collaboration website: www.ligo.org. For a detailed overview of the science of gravitational wave instrumentation, see Saulson (2017).

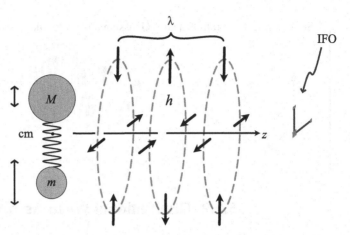

Figure 5.5 A simple gravitational wave emitter, the strain field imposed on space by the passing gravitational wave, and a Michelson interferometer (IFO) placed in the path of the wave to detect the strain field. The size of the interferometer in comparison to a wavelength is greatly exaggerated in this diagram.

changes. Photodiodes at the antisymmetric output port of the interferometer record the optical power change and thereby the gravitational wave strain.

Gravitational waves detected by current interferometric gravitational wave detectors have oscillation periods in the millisecond range. In order to maximize the sensitivity of interferometers to such waves, we would ideally want arms long enough that light takes half a period to traverse an arm. That way, the effect of the gravitational wave on the phase of the light returning from the arm is maximum. Since gravitational waves and light waves travel at the same speed, c, the interferometer arms would ideally be half a wavelength long, something like 500 km! Unfortunately, an interferometer of such length is prohibitively expensive to build on earth and instead the arms are a "modest" 4 km long but have optical cavities in them to increase the light storage time. (Optical cavities are covered in detail in Chapter 7.) In fact, light attempting to leave the interferometer is also redirected back in through the use of optical cavity mirrors at the input (symmetric) and output (antisymmetric) ports of the interferometer. Eventually, after what is sufficient storage time in the interferometer, light gets out, with luck having accumulated an extra phase shift difference from a passing gravitational wave. Figure 5.6 shows the layout of a modern gravitational wave interferometer. Gravitational wave detectors based on Michelson interferometers are most sensitive to gravitational waves incident from directly above or below (provided the polarization of the wave is aligned with the detector arms). This is due to a fortunate match between the shape of the strain field induced by the wave and the interferometer itself. Conversely, if the wave is incident in the plane of the arms and at 45° between the arms, the detector is insensitive to it regardless of polarization. We really are using a Michelson interferometer as an antenna. So, like other antennas it has an antenna pattern describing this sensitivity variation with direction. The antenna pattern of a Michelson used in this way is shown in Figure 5.7.

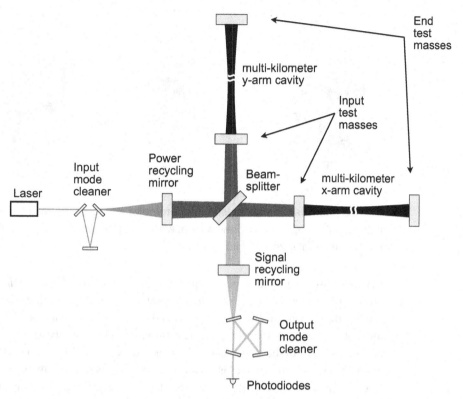

Figure 5.6 The basic layout of a modern gravitational wave interferometer. (Not to scale.) The shade of light is a guide to the optical power in each section. The darker the shade, the more light. The arms have by far the most circulating power of any part of the system. Even though the laser power is only on the order of tens of Watts, the arm cavity power can reach hundreds of kiloWatts due to the long light storage time in the arm cavities. The laser is actually an impressive system in and of itself, incorporating numerous noise-reduction and optical improvement stages not shown here. The input mode cleaner can be considered the last stage of the laser and serves to present the interferometer with an exceedingly pure TEM_{00} mode. The power recycling mirror increases the circulating power in the interferometer by redirecting light back into the interferometer. Light would otherwise be lost out the symmetric port. The signal recycling mirror at the antisymmetric port allows the response of the interferometer to different gravitational wave frequencies to be adjusted so as to take full advantage of the low-noise parts of the signal band. (Without the signal recycling mirror, the interferometer response would be dictated by the light storage time in the arms.) The output mode cleaner's role is to separate the optical mode containing the gravitational wave signal from optical modes containing only noise or control signals.

The big challenge in detecting gravitational waves with interferometers or indeed any other way, is the extremely small signal they present. So, it's necessary to use an instrument that picks up as much signal as possible (signal power and bandwidth) while having

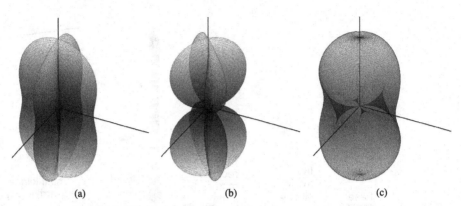

(a)   (b)   (c)

Figure 5.7 Antenna patterns for a Michelson interferometer acting as a gravitational wave detector: **(a)** "+"-polarized gravitational waves, **(b)** "×"-polarized gravitational waves, **(c)** randomly polarized gravitational waves.

very low noise. Other things being equal, this favors broadband instruments. Current interferometric gravitational wave detectors' signal band covers about three decades starting at a few Hertz and reaching up to a few thousand Hertz. (The "signal band" of a detector is the range of frequencies over which the detector has good sensitivity to signals.) LIGO's signal band is much of the same band as humans can hear and the conversion of gravitational waves directly to sound waves is a very nice way of experiencing them.

At the time of writing, there are three large scale gravitational wave interferometers doing astronomical observations and more are being built. These interferometers are operated as a network rather than individually. Using several gravitational wave interferometers together boosts the overall signal-to-noise ratio and gives immunity to false signals and glitches (since one expects a proper gravitational wave to show up in all detectors at about the same time). Also, since the antenna pattern of each detector is rather omni-directional, three detectors are needed to obtain an accurate sky location of any gravitational wave source.

Exercises

5.1 The output of a Michelson interferometer shows four bright vertical fringes across the recombined beam. Assume the beam has an approximate diameter of 2 mm. Assume the fringes are due to angular misalignment between the two beams, one from each arm, which superpose to form the output beam. Is the misalignment in pitch or yaw? Estimate the magnitude of the angular misalignment in degrees.

5.2 Figure 5.1a shows that in a Michelson interferometer with equal-length arms, the beams on the beamsplitter do not need to be recombined in the same place they were split. Using sketches or otherwise, show that it is not possible to perfectly align a Michelson whose arms are *unequal* in length unless all beams are incident on the beamsplitter at a single point. *Hint*: Show that the two beams coming from the arms cannot be recombined at the beamsplitter in such a way as to make them overlap perfectly and be perfectly parallel.

5.3   The field reflected at a beamsplitter obtains a minus sign upon reflection from one direction but not the other. If neither or both reflections obtained a minus sign, show that energy would not be conserved in a Michelson interferometer.

5.4   Figure 5.9 shows the interference between a beam with spherical wavefronts and a beam with approximately flat wavefronts. Derive an expression for the radii of the rings. *Hint*: The overall phase difference between the beams determines the radius of the innermost ring.

5.5   Could you build a Michelson interferometer with three arms at right angles to one another, one along each coordinate axis? If one uses conventional beamsplitters that split a single beam into only two beams, one would need to use two beamsplitters. Assume deviations away from the center of the fringe are small. Describe how the signals at the outputs of the two beamsplitters should be combined to obtain length difference signals between any two of the three arms. *Hint*: A beamsplitter has two outputs, one of which normally travels back toward the laser. Assume it is possible to access both outputs (perhaps using optical isolators).

5.6   Consider a Sagnac interferometer like the one in Figure 5.1b forming a square of side L with the beamsplitter at one vertex and mirrors at each of the other three vertices. When the Sagnac is stationary, the recombined beam exiting toward the photodiode experiences destructive interference and so the photodiode receives no light. Calculate the speed at which the interferometer must spin (in the plane of the beams) in order to maximize the brightness at the photodiode. (Assume non-relativistic speeds.) *Hint*: The mirror faces are tangent to the direction of motion, while the beamsplitter's reflective surface is perpendicular to the direction of motion.

5.7   One wavefront normal passing through and reflecting within a Fabri–Perot Etalon is shown in the figure herein.

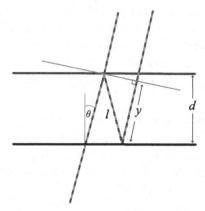

By considering the parallel rays leaving the etalon, explain why the following condition for constructive interference is correct.

$$2l - y = n\lambda, \qquad n \in \mathbb{Z}.$$

The ray travels a distance l from one plate to the other. By finding an expression for the distance y, show that the condition for constructive interference can be written

$$1 - \sin^2\theta = \frac{n_0 - m}{2l}\lambda,$$

where $m = 1, 2, \ldots$ and n_0 is the number of wavelengths corresponding to the shortest round-trip distance between the plates, rounded *up* to the nearest integer. In other words

$$n_0\lambda = 2d + \Delta\lambda,$$

where $\Delta\lambda$ is the part of a wavelength by which $n_0\lambda$ exceeds $2d$. Assuming $\theta \ll 1$ and $\lambda \ll d$, show that bright fringes occur for wavefront normals with angles of incidence

$$\theta_m = \sqrt{\frac{m\lambda - \Delta\lambda}{2d}}.$$

5.8 (Computer problem) A high-reflective (HR), quarter-wave stack mirror is made from a stack of dielectric layer pairs. Each layer pair consists of a low-index layer and a high-index layer each of which has an optical thickness (actual thickness times the index of refraction) equal to a quarter wavelength of the light the coating is designed to reflect. Plot the reflectance at normal incidence, $R = r^2$ as a function of wavelength, 700 nm $\leq \lambda <$ 1350 nm, of a 15 layer-pair, quarter-wave stack where the high-index layer is tantalum pentoxide with index of refraction $n_H = 2.2$ and the low-index layer is fused silica with index of refraction $n_L = 1.46$. The design-wavelength (center wavelength) of the coating is $\lambda_0 = 1064$ nm. Compare your result with the measurement shown in Figure 5.3b.

5.9 (Computer problem) Plot the interferogram output of a Fourier-transform spectrometer when the input beam consists of light emitted from a sodium lamp filtered around the doublet at 589.0 nm and 589.6 nm. Assume these are the only wavelengths present in the spectrum and they have equal irradiance. Plot 100,000 points and assume the mirror moves by a distance $\epsilon = 1$ mm. *Hint*: No integrals are required. What is the minimum distance the mirror must move to resolve the two lines?

5.10 A "perfect optical band-pass filter" would admit all light between wavenumbers $k_1 = k_0 - \frac{\Delta k}{2}$ and $k_2 = k_0 + \frac{\Delta k}{2}$ and extinguish light of any other wavenumber. Show that the interferogram for such filtered light would be $I_{\text{det}} = \frac{I_{\text{in}}}{2}\left[1 - \cos(k_0 x)\operatorname{sinc}\left(\frac{\Delta k}{2}x\right)\right]$. Show that as the passband of the filter tends to zero, we regain the expected interferogram of monochromatic light, Eq. (5.16).

5.11 (Computer problem) The interferogram output of a Fourier-transform spectrometer has the form of a damped harmonic oscillator

$$I_{\text{det}}/I_{\text{in}} = \frac{1}{2} + e^{-x/x_e} \cos(k_0 x),$$

where $k_0 = \frac{2\pi}{\lambda_0}$, $\lambda_0 = 1$ μm, and $x_e = 1 \times 10^{-5}$ m. Either numerically or analytically, find and plot the corresponding spectrum $f(k)$. It should have a peak of finite width centered at k_0. The peak shape is known as a Lorentzian. It's typical of spectral lines and is the frequency response of a damped harmonic oscillator.

5.12 By considering the antenna polarization in Figure 5.7c, use sketches to argue that there is a selection effect so that gravitational waves from the region of the sky near the north celestial pole will be underrepresented. Current detectors all lie at latitudes between 30°N and 46°N. *Hint*: Consider first the case where all detectors lie at the equator.

5.3 Experiment: The Michelson Interferometer

Objectives

1 Learn to set up and align a Michelson interferometer.

2 Measure the index of refraction of glass with 1% accuracy or better.

3 Derive the expected fringe spacing for Newton's rings and compare with observation.

Equipment Needed

- Helium-Neon laser.
- Non-polarizing cube beamsplitter plus mount.
- Small piece of plate-glass, 1–3 mm thick, 10–50 mm in diameter, placed in a mount that allows for very fine yaw adjustment via thumbscrew.
- Photodiode with photodiode amplifier.
- Real-time oscilloscope (such as a traditional analog oscilloscope).
- *Optional*: A small amplified loudspeaker with volume control and a converter from photodiode output to the loudspeaker.

Understanding the Interferometer

Alignment

Set up a Michelson interferometer and obtain interference fringes. You will need two adjustable mirrors between the laser and the beamsplitter that allow you to adjust the input beam to the interferometer. To change the angle at which the beam enters the interferometer, use the downstream mirror. To translate the beam laterally, use the upstream mirror to move the beam on the downstream mirror, then use the downstream mirror to correct the pitch or yaw so that a pure translation results. Once the interferometer is roughly aligned, you will want to expand the output beam on the projection screen.

> *Recommendations*: Table-style cube beamsplitter mounts are common and very easy to use. However, since the beamsplitter is likely to be stored unmounted, be careful only to place it down on the sides that are ground glass. Hold it by the corners with gloved hands and don't put it down on any of the four polished sides.

Align the interferometer so that you see a number of straight vertical fringes across the screen. This is sometimes called the "classical alignment." With this alignment, the wave fronts from each arm are incident on the same location on the projection screen but come in at slightly different angles. In other words, you've actually intentionally slightly misaligned the beams coming from the two arms. If you press lightly on the posts supporting the endmirrors of the arms, you should be able to see rapid fringe motion. How does the direction of motion of the fringes depend on which arm is being shortened/lengthened? Explain why the fringes move in the observed direction for a given choice of arm and direction of mirror motion.

Linear Superposition

Explain how the fringes are produced. Where are they produced, at the viewing screen, in the beamsplitter, on your retina? While you are considering this question, it may help to remember that the fields from the two beams add linearly. Now work to align the interferometer so that the projected spot just blinks when an arm length is changed. In other words, align it well enough that the fringe size becomes larger than the spot size. We will monitor the light level with a photodiode. This is the most common mode of operating a Michelson interferometer nowadays. With the interferometer aligned in this fashion, detect the output light on an amplified photodiode and observe the signal on an oscilloscope. Press lightly on one of the mirror mounts so that it is briefly placed into forward motion due to the compliance of the mounting post. While the mirror is in forward motion, you should see a sine wave on the oscilloscope. This sine wave demonstrates that the electric (and magnetic) fields in the two beams are themselves sinusoidal. Starting with results derived in the theoretical section of this chapter, consider whether this conclusion is unavoidable.

If your instructor has provided a loudspeaker for listening to the output of your interferometer, consult with your instructor before connecting it. **Do not put the loudspeaker near your ears and make sure sound level is safe before proceeding.** Repeat the above-mentioned process while listening to the output of the interferometers. Tap on some of the components. Describe and explain where the different sounds are coming from. Does feedback occur? If so, what is the most likely path?

> *Recommendations*: Make sure that the output beam is not larger than the active area of the photodiode. Use a focusing lens if necessary. Also, be sure that the light level is never large enough to saturate the photodiode amplifier, or use a lower-gain photodiode amplifier.

Index of Refraction Measurement

Put the interferometer back into the classical alignment so that you see about five vertical fringes across the beam. You will now use the sensitivity of the instrument to optical path length variation between the arms to measure the index of refraction of glass with better than 1% accuracy. Obtain an appropriate small piece of plate-glass to insert into one of the arms. Before mounting and inserting the glass into the arm, read this entire section carefully and perform all the required calculations.

Once the glass is inserted into the beam, rotating the glass about the vertical will change the angle of the glass with respect to the beam. As the glass is rotated you should see fringes moving, a.k.a. "wrapping," across the screen. The reason for this is that changing the angle of the glass plate w.r.t. the beam changes the amount of glass through which the beam must travel. The index of refraction of glass is higher than of air, this increases the effective pathlength, known as the "optical path." (Since the frequency of the light doesn't change as it passes through the material, the reduced speed of light in the medium implies that the wavelength is reduced, or else $v = f\lambda$ would be violated.) The optical path is $l_{opt} = n \times l_{physical}$ where $l_{physical}$ is the actual physical path length, as measured by a ruler. The increased optical path implies that the beam that traverses the glass accumulates additional phase with respect to the other arm. The pathlength in the glass increases as the glass is rotated away from normal incidence so the phase accumulation also increases. With the glass at about 45° to the beam, you will be observing the fringe motion as the glass is rotated slightly. By counting fringes as a function of rotation angle, you will measure the index of refraction of the glass plate very precisely.

The easiest way to measure the rotation angle is by monitoring the motion of the spot reflected from the glass. (Since the glass is at about 45°, the reflected spot is about perpendicular to the arm and can be viewed on a screen or wall.) Note that the reflected spot moves through twice the angle of the plate glass itself. Starting at $\theta_0 = 45°$, carefully rotate the glass through a small angle $\Delta\theta$ while counting the number of fringes ΔN going by at the output of the interferometer. Repeat the measurement several times and make a table of ΔN versus $\Delta\theta$. The number of fringes passing by ΔN is related to the change in the optical path of the arm by

$$\Delta N = \frac{2\Delta l_{\text{opt}}}{\lambda}. \tag{5.32}$$

If you are using a red HeNe laser for this experiment, then $\lambda = 632.8$ nm. The factor of two in this equation arises from the fact that each photon sees the optical path length change twice, once on the way toward the end mirror of the arm and once on the way back to the beamsplitter. Equation (5.32) implies that

$$\Delta N = m(n, \theta_0)\Delta\theta, \tag{5.33}$$

where

$$m(n, \theta_0) \equiv \frac{2}{\lambda} \left. \frac{d l_{\text{opt}}}{d\theta} \right|_{\theta_0} \tag{5.34}$$

is the slope of the data ΔN vs. $\Delta\theta$. The slope depends on both the index of refraction n of the glass plate and the center angle of incidence θ_0. You will fit your data, ΔN versus $\Delta\theta$, to a straight line and obtain a measured slope for $m(n, 45°)$. By comparing the measured slope to the function $m(n, 45°)$, you will obtain the value of n. Beware, $m(n, \theta_0)$ is not a simple function. See the recommendations mentioned later before trying to calculate it. It's also possible to find $m(n, 45°)$ numerically and solve the equation $\frac{\Delta N}{\Delta\theta} = m(n, 45°)$ graphically (or numerically).

Make sure you find the uncertainty in your measured m by supplying the fitting routine with appropriate uncertainties $\Delta\theta$ and ΔN. Use the uncertainties in m to find the uncertainty in n. You need to come up with experimental procedures that will allow you to measure $\Delta\theta$ and ΔN sufficiently accurately to find n to 1% or better. You will need to consider how to mount the plate glass so that you can rotate it sufficiently smoothly and slowly so as not to lose track of fringes during the measurement. You will also need to measure $\Delta\theta$ with sufficient precision.

Recommendations: The full optical path length taken by a beam as it travels once between the mirror and beamsplitter is

$$l_{\text{opt}} = l_{\text{air}} + n l_{\text{g}}, \tag{5.35}$$

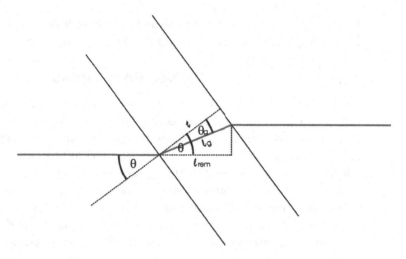

Figure 5.8 Schematic representation of the path of a beam of light through a piece of plate glass.

where l_{air} is the physical length of the path the light travels in air and l_g is the physical length of the path that the light travels inside the glass. Due to Snell's Law, the path inside the glass is at an angle $\theta - \theta_g$ with respect to the path that would have been taken by the beam if the glass were not present (see Figure 5.8). The component l_{rem} of the in-glass path that is parallel to the in-air path is the length of the path that would have been traversed in air had the glass not been present. In other words, the total in-air path is now only $l_{air} = l_{arm} - l_{rem}$ where l_{arm} is the length of the arm, and l_{rem} is the path length removed given by the projection of the in-glass path onto the axis of the in-air beam. The distance traveled by the beam inside the glass is

$$l_g = t/\cos\theta_g. \tag{5.36}$$

The total optical path between the beamsplitter and end-mirror is then

$$\begin{aligned} l_{opt} &= l_{air} + n l_g \\ &= l_{arm} - l_{rem} + \frac{nt}{\cos\theta_g}. \end{aligned} \tag{5.37}$$

The arm length l_{arm} is constant, so

$$\frac{d l_{opt}}{d\theta} = -\frac{d l_{rem}}{d\theta} + \frac{d}{d\theta}\left(\frac{nt}{\cos\theta_g}\right). \tag{5.38}$$

Evaluate this derivative at θ_0 and use it to obtain $m(n, \theta_0)$ from Eq. (5.34).

When you fit your data to a straight line, you will need to incorporate uncertainties in both ΔN and $\Delta\theta$. One way of dealing with simultaneous horizontal and vertical error bars in a straight-line fit is to make a preliminary fit without uncertainties. Use the slope of this preliminary fit to convert the horizontal error bars to equivalent vertical error bars. Then add these equivalent vertical error bars *in quadrature* to the original vertical error bars to obtain total effective vertical error bars. Perform the final fit using the total effective vertical error bars and no horizontal error bars. This approach allows you to find the parameter uncertainties with a direct application of the method in Section 2.3.2.

Curved Phasefronts

Replace both end-mirrors of the interferometer with beam blocks and replace the glass plate with a plano-convex lens oriented so that its optic axis is perfectly aligned with the beam and with the planar side facing the beamsplitter. After careful alignment, you should see circular fringes at the output of the interferometer like the ones shown in the figure below. The fringes are due to the interference between the beams reflected from the two sides of the lens, one flat, the other curved. Measure and explain the characteristic spacing of the fringes as a function of radial distance. These types of fringes are known as "Newton's rings." Extract a vertical cut through the center of the pattern and show that the envelope of the fringes has the expected dependence on distance from optic axis.

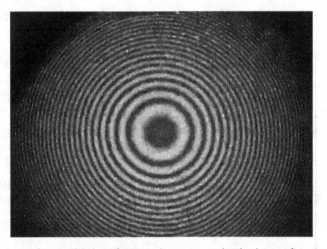

Figure 5.9   Newton's rings are due to the interference between spherical wavefronts with different curvature. In the case shown here, only one of the beams has wavefronts with significant curvature while the other beam has wavefronts that are approximately flat.

Ideas for Further Investigation

Take a photo of the vertical fringes of a Michelson interferometer due to slight horizontal misalignment of an end mirror. Extract a cut that shows the fringe irradiance as a function of horizontal position. From first principles, calculate the expected functional form of the irradiance along the cut and compare (fit) it to your result.

Estimate the coherence time in nanoseconds of the laser you're using by attempting to get interference with increasingly mismatched arm lengths.

Lasers

6.1 How Lasers Work

Lasers are light sources whose outstanding characteristics are high spatial and temporal coherence and the associated ability to form diffraction limited beams. These features make lasers useful in applications as diverse as cutting steel, measuring the distance to the moon, cooling atoms to near absolute zero, and detecting gravitational waves. These same features mean that many lasers can cause permanent damage to the eye and precautions to prevent injury must be taken when working around lasers (see Section 2.1).

Laser light is generated in transitions from higher energy quantum states of atoms or molecules to lower energy ones. Thus, lasers require some "active medium" within which the participating atoms/molecules reside. The fixed energy difference between quantum states accounts for the highly monochromatic nature of the emitted light. (Recall that the energy carried by a photon is related to its frequency by Plank's constant $E = h\nu$.) In lasers, we arrange for a majority of the participating atoms/molecules to be excited, a situation known as a "population inversion." As these excited states decay, they emit laser light. A source of power is required to re-excite the atoms/molecules in order to maintain the population inversion – a process known as "pumping."

The active medium and the pumping method vary enormously between lasers. In a HeNe gas laser, the active medium is the helium and neon gas mixture and the population inversion is achieved by applying a high voltage across the gas, forcing an electric current to travel through it. In a neodymium-YAG solid state laser, the active medium consists of neodymium atoms scattered throughout a transparent YAG crystal (yttrium aluminum garnet). In this case, it's the neodymium atoms that produce the light and the system is optically pumped using a flashlamp or laser diodes. Several other types of lasers exist, for example, diode lasers, dye lasers, chemical lasers, and so forth, each with their own active medium and pumping system.

6.1.1 Stimulated Emission

Lasers produce *coherent* emission due to the process of stimulated emission, which defines the operation of a laser.[1] Consider an active medium that has been pumped so that it now has a majority of participating atoms in an excited state. (For the purpose of illustration,

---

[1] The word laser was originally conceived as an acronym for Light Amplification by Stimulated Emission of Radiation. However, "laser" is now considered a word in its own right.

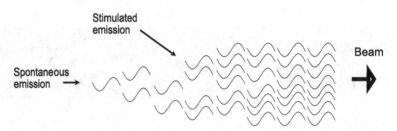

The stimulated emission process builds up a significant beam. Not every photon stimulates an atom to emit another photon but if the population inversion is sufficiently large, a beam will build up as shown.

we'll assume the relevant energy levels are atomic energy levels as opposed to molecular ones, etc.) An excited atom decays spontaneously and emits a photon in the process. The photon doesn't travel far before encountering another excited atom that it now *stimulates* to decay causing the release of a second photon. The second photon has the same frequency, phase, and direction of motion as the original photon. Now there are two photons that are in phase and traveling in the same direction. Each of these photons stimulates emission from another atom, leading to four coherent photons traveling in the same direction, and so on. Not every photon stimulates the emission of another photon; some photons are absorbed or scattered. However, as long as the population inversion is high enough and the optical losses low enough, the stimulated emission process will build up a strong coherent beam (see Figure 6.1).

From the description so far it would seem that a pumped medium might radiate in an arbitrary direction since the direction of the initial spontaneous emission would be random. In practice, in order to get an appreciable buildup of optical power, photons leaving the active medium are "recycled" by redirecting them back into the medium so that they can stimulate more emission. The active medium is placed between two mirrors that face one another forming the "laser cavity." When the mirrors are correctly aligned so that any given photon travels along the same back-and-forth path a large number of times, the emission process becomes efficient enough that a bright beam appears within the cavity. The direction of light travel is therefore always along the cavity axis. Typically, one of the cavity mirrors is chosen to be slightly transmissive so that cavity-light can leak out and form the output beam of the laser (see Figure 6.2).

6.1.2 The Two-level Model

A full description of the interaction of light and matter in lasers involves statistical mechanics, electrodynamics, and quantum mechanics. Due to this richness, laser physics is a significant subfield of physics and numerous large textbooks have been written on the subject of lasers alone. Here we'll introduce the simplest mathematical model for the lasing process – a simplified two-level model.[2] We consider the transitions of electrons between an atomic ground state with energy E_1 and an excited state with energy $E_2 > E_1$. We

---

[2] The treatment here uses the same notation as popular laser texts such as Siegman (1986) and Svelto (2010), any of which can be consulted for more details.

Figure 6.2 Photograph of an open cavity helium-neon laser. The glow from the tube is pink and contains the emission from all the atomic transitions excited by the electrical discharge traveling through the tube. The system is lasing on a single neon transition, as can be seen from the spots on the cavity mirrors and also on the Brewster-angle windows on each end of the tube. The laser emission is bright red at 632.8 nm. The right-hand mirror is slightly transmissive, so the output of this laser travels off to the right in this photograph.

assume that these are the only states available and at any given time there are N_1 atoms per unit volume in the ground state and N_2 atoms per unit volume in excited state. At thermal equilibrium at temperature T, the average population numbers are given, according to statistical mechanics, by Maxwell–Boltzmann statistics.

$$N_1 = C \exp\left(\frac{-E_1}{k_B T}\right)$$
$$N_2 = C \exp\left(\frac{-E_2}{k_B T}\right),$$

(6.1)

where C is a unitless constant. This implies that the ratio of the excited population to the unexcited population is

$$\frac{N_2}{N_1} = \exp\left[\frac{-(E_2 - E_1)}{k_B T}\right].$$

(6.2)

Note that since $E_2 > E_1$, we have $N_2/N_1 < 1$. In other words, the majority of atoms are unexcited at thermal equilibrium. Since photons emitted from the transition $2 \rightarrow 1$ have energy $hv = E_2 - E_1$ (where v is the frequency of the light and h is Planck's constant) we can also write the ratio in terms of the emission frequency

$$\frac{N_2}{N_1} = \exp\left(\frac{-hv}{k_B T}\right). \tag{6.3}$$

Since $N_1 > N_2$ at thermal equilibrium, we will need to pump to obtain the necessary population inversion. We'll discuss the pumping process shortly. For now, imagine that we've succeeded in obtaining a population inversion and we have a cavity beam with frequency v and irradiance I traveling within the active medium. The atoms in the active medium are bathed in this irradiance, which continuously triggers stimulated emission from excited atoms but is also absorbed by unexcited atoms. Whether the beam grows in strength will depend on whether stimulated emission outpaces the absorption. In any given unit of time, each excited atom will have some probability p_{st} of being stimulated to emit a photon. Similarly, each unexcited atom will have some probability per unit time p_{ab} of absorbing a photon from the irradiance. Each of these probabilities will depend on the strength of the irradiance I in which the atoms are bathed according to

$$p_{st} = \sigma_{21}\left(\frac{I}{hv}\right), \tag{6.4}$$

$$p_{ab} = \sigma_{12}\left(\frac{I}{hv}\right). \tag{6.5}$$

Here I/hv is the photon flux in the beam, defined as the number of photons passing through a unit perpendicular area per unit time. The σ's are the so-called cross-sections for the respective processes. The cross section σ_{21} is just the probability that a given atom will be stimulated to emit a photon in any given unit of time *per* unit of photon flux bathing that atom.[3] Similarly, σ_{12} is the probability that a given atom will absorb a photon in any given unit of time, per unit photon flux. If an atom in the upper state is stimulated to emit, it undergoes the transition $2 \to 1$, so N_2 is reduced by one while at the same time N_1 is increased by one. The rate at which the upper- and lower-level occupancies change are then

$$\frac{dN_2}{dt} = -p_{st}N_2 + p_{ab}N_1 = (-\sigma_{21}N_2 + \sigma_{12}N_1)\frac{I}{hv} \tag{6.6}$$

$$\frac{dN_1}{dt} = p_{st}N_2 - P_{ab}N_1 = (\sigma_{21}N_2 - \sigma_{12}N_1)\frac{I}{hv}. \tag{6.7}$$

Due to time-reversal symmetry, the $2 \leftrightarrow 1$ transition can run in either direction and absorption and emission have the same cross-sections: $\sigma_{21} = \sigma_{12} \equiv \sigma$. The "rate equations" become

$$\frac{dN_2}{dt} = -(N_2 - N_1)\frac{\sigma I}{hv} \tag{6.8}$$

$$\frac{dN_1}{dt} = (N_2 - N_1)\frac{\sigma I}{hv}. \tag{6.9}$$

---

[3] σ_{21} has units of area, and hence the name. Actually, if the excited atoms were marbles and the photons in the flux were infinitesimally small particles traveling along parallel paths that were randomly distributed in space, and if any marble that is hit will emit, then σ_{21} would be the actual projected area of the excited atoms. This rather inaccurate picture of the situation is sometimes called the "buckshot model." Mainly it helps one to understand why σ has units of area.

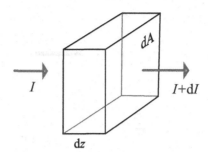

Figure 6.3   Slice of active medium bathed in the irradiance of a beam.

Basically, they say that whichever level has more occupancy will be depleted. So, if we have a population inversion $N_2 > N_1$ then the upper level will be depleted by stimulated emission and in the process, emit one photon for every reduction in N_2 by one. (Clearly, we would need to pump to maintain $N_2 > N_1$ but a pumping term is not included in these simplified rate equations.)

The net rate at which photons are generated per unit volume is equal to the rate of stimulated emission per unit volume

$$\frac{\mathrm{d}n}{\mathrm{d}t} = (N_2 - N_1)\frac{\sigma I}{h\nu}. \tag{6.10}$$

The net optical power increase in some infinitesimal volume $\mathrm{d}V$ is

$$\mathrm{d}P = h\nu\frac{\mathrm{d}n}{\mathrm{d}t}\mathrm{d}V. \tag{6.11}$$

We will choose our axes so that the z-axis lies along the beam's direction of propagation within the active medium. We consider a volume slice $\mathrm{d}V = \mathrm{d}A\mathrm{d}z$ (see Figure 6.3) through which the beam is passing. The increase in irradiance as the beam passes through the volume is just the increase in optical power per unit perpendicular area, $\mathrm{d}I = \frac{\mathrm{d}P}{\mathrm{d}A}$. The change in irradiance per unit propagation distance $\mathrm{d}z$ is then $\frac{\mathrm{d}I}{\mathrm{d}z} = \frac{\mathrm{d}P}{\mathrm{d}A\mathrm{d}z} = \frac{\mathrm{d}P}{\mathrm{d}V}$. Substituting into Eq. (6.11) from Eq. (6.10) gives

$$\frac{\mathrm{d}I}{\mathrm{d}z} = (N_2 - N_1)\sigma I. \tag{6.12}$$

The active medium has net positive gain and lasing will occur if $\frac{\mathrm{d}I}{\mathrm{d}z} > 0$. Gain is positive whenever we have a population inversion: $N_2 > N_1$. The trick is to ensure that the population inversion is maintained by pumping.

Unfortunately, an actual two-level laser doesn't work because a population inversion can't actually be achieved. The only way to excite the lower level is by bathing it in a photon flux of frequency ν. Equation (6.9) gives the rate at which such light is absorbed and implies that once $N_1 = N_2$, absorption will stop. When $N_1 = N_2$, the medium is neither absorptive nor does it have gain. Any attempt to pump it results in emission identical to any pump light absorbed and no net energy is deposited in the active medium. In effect, the material becomes transparent to the pump light.

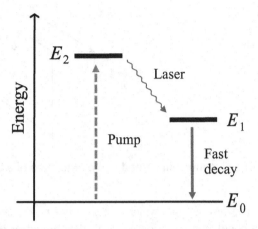

Energy-level diagram for a typical three-level laser. In this diagram the upper transition corresponds to stimulated emission while the lower one corresponds to rapid but unstimulated decay, typically non-radiative. In other lasers, such as the original ruby laser, this is reversed; the upper transition being the rapid unstimulated decay and the lower transition being the laser transition.

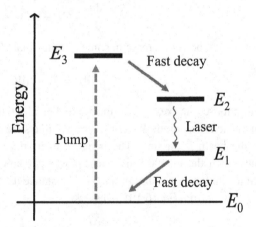

Energy-level diagram for a typical four-level laser.

The trick is to use more than two energy levels. The pumping process excites the atoms to some high energy level and stimulated emission occurs for *just one* of several transitions required to reach the ground state. (The remaining decays occur spontaneously.) That way the pumping process doesn't rely on the exact same mechanism as the emission process and a population inversion can be achieved. The simplest version of such a laser is a three-level laser whose energy states are shown in Figure 6.4. In both three-level and four-level lasers the stimulated emission occurs between two states for which a population inversion can be efficiently arranged. The energy levels of a typical four-level laser are shown in Figure 6.5.

6.1.3 A Four-level Laser: The HeNe

A HeNe (pronounced "hee-nee"), or helium-neon laser is one of the most recognizable lasers available. The laser tube is cylindrical and the enclosure is often cylindrical to match. If the power supply is incorporated in the enclosure, then it usually has the shape of a rectangular cylinder. HeNes were invented in the early 1960s but are still an enormously popular choice for laboratories because they possess the best optical properties of any laser you can buy for the price of a modest computer. A typical HeNe has a long coherence length, a few tens of cm for mass-produced versions and much longer for frequency-stabilized models. The range of frequencies emitted, the linewidth, is typically about 1 GHz, which is very narrow compared to say, a basic diode laser whose linewidth may be 1,000 times larger. Frequency-stabilized (single-mode) HeNes do a thousand times better again, with linewidths around 1 MHz.

The helium-neon laser is categorized as a four-level laser because the neon atoms that emit the actual laser light have four electron energy levels that take part in the lasing process. The helium atoms on the other hand participate in the pumping process. They are used to absorb the pump energy from an electrical discharge sent through the helium and neon gas mixture. The helium atom happens to have a pair of energy levels whose energies, E_3 and E_5, coincide remarkably closely to the energies of a pair of neon levels (see Figure 6.6). If a helium atom has excited electrons in these levels, then their energy can be transferred directly to electrons in a neon atom whenever the two atoms approach sufficiently closely or collide. This energy exchange makes the pumping process much more efficient than trying to pump the neon levels directly. The E_3 and E_5 levels of helium are very long-lived states and are good at accumulating pump energy. Helium electrons excited by the discharge to all levels of energy higher than E_3 and E_5, tend to decay down to this pair of levels and then sit there for a long time.[4] The E_3 and E_5 in helium act as "holding levels" that gather excited electrons from many higher energy decays. During the period in which a particular helium atom's E_3 and E_5 levels are excited, it has a significant chance of encountering a neon atom and transferring its energy to it. Thus, a much larger fraction of neon atoms gets excited to E_3 and E_5 than would occur in the absence of helium. The upper levels of the neon atom can decay via stimulated emission to a number of states. The resulting transitions and their associated photon wavelengths correspond to the different possible colors of HeNe lasers.

Since the E_5 and E_3 levels of neon are being pumped efficiently, a population inversion can occur for laser transitions $5 \rightarrow 4$, $5 \rightarrow 2$, and $3 \rightarrow 2$. To maintain such a population inversion, we need E_2 and E_4 to get emptied rapidly whenever they get filled in order to maintain low occupancy. They're able to do so by decaying to the E_1 level through rapid spontaneous emission. To maintain good drainage of the lower levels, the E_1 level also needs to decay quickly to the ground state, E_0. The decay $1 \rightarrow 0$ occurs due to neon atom collisions with the glass tube containing the gas. The neon atoms transfer energy to the walls of the tube, which must be very close for this process to happen sufficiently quickly.

---

[4] That's because the transitions out of these levels are "forbidden transitions." The required transitions have a hard time radiating because no electric dipole moment change is induced during the transition. So, the levels don't decay very efficiently.

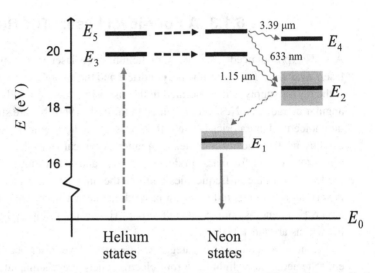

Figure 6.6 Energy-level diagram for the helium-neon lasing process. The light gray dashed arrow represents the pump. The black dashed arrow represents energy transfer from helium to neon via atomic collisions. The solid arrow corresponding to the decay from E_1 to the ground state, E_0, represents atomic collisions with the walls of the capillary in the laser tube. The wiggly dashed line $E_2 \rightarrow E_1$ represents fast spontaneous decay. (A similar decay should be indicated $E_4 \rightarrow E_1$ but has been omitted to prevent clutter.) The solid wiggly lines represent stimulated emission. Three emission lines of the HeNe are shown. The gray boxes surrounding E_1 and E_2 represent a set of nearby levels that can all play a similar functional role in the lasing process. For example, HeNe emission wavelengths beyond those shown in the diagram may be obtained by using a different one of the E_2 lines as the lower lasing level.

That's why all HeNe lasers have a narrow capillary along the axis of the laser tube. It's only inside this capillary that lasing can happen.

6.1.4 Rate Equations

Rate equations, like Eqs. (6.8) and (6.9), are models for the dynamic behavior of a laser. They are sets of coupled first-order differential equations that describe the rates at which occupancy of the various energy levels changes. The rate equations in Section 6.1.2 confirmed our basic understanding that the gain in an active medium, $\frac{dI}{dz} > 0$, depends on the population inversion. In this section, we consider the rate equations for a three-level laser where the lasing transition occurs between the uppermost and middle levels, as in Figure 6.4.

The rates $\frac{dN_i}{dt}$, ($i = 0, 1, 2$) for the occupancy of the three levels are

$$\frac{dN_2}{dt} = R_p - \gamma_{21}N_2 - \sigma_{21}N_2\left(\frac{I}{h\nu}\right) + \sigma_{12}N_1\left(\frac{I}{h\nu}\right) \tag{6.13}$$

$$\frac{\mathrm{d}N_1}{\mathrm{d}t} = -\gamma_{10}N_1 + \gamma_{21}N_2 + \sigma_{21}N_2\left(\frac{I}{h\nu}\right) - \sigma_{12}N_1\left(\frac{I}{h\nu}\right) \tag{6.14}$$

$$\frac{\mathrm{d}N_0}{\mathrm{d}t} = -R_p + \gamma_{10}N_1. \tag{6.15}$$

The first equation describes the occupancy changes of the uppermost level. The first term is the pumping rate R_p; it's the number of atoms excited to the energy level E_2 per unit time, by whatever pumping process is being used. R_p is a constant that depends on the pump mechanism and pump power. We can't pump hard enough to fully deplete the ground state, so there are always unexcited atoms available to be pumped and the pump rate is independent of the occupancy of the ground state, N_0. The next term, $\gamma_{21}N_2$, is the rate of spontaneous transition $2 \rightarrow 1$. γ_{21} is the probability that any given atom will spontaneously transition $2 \rightarrow 1$ in a unit of time. The third term describes the rate of stimulated emission and corresponding transition $2 \rightarrow 1$. It is proportional to both the occupancy N_2 and the photon flux $\frac{I}{h\nu}$. The fourth term describes absorption of laser photons by the lower lasing level with corresponding transition $1 \rightarrow 2$. In the second equation, the first term describes spontaneous decay to the ground state. (There is no direct pumping to E_1, so there's no pump term present.) Apart from that, the second equation only contains terms corresponding to the $1 \leftrightarrow 2$ transitions already discussed. The last equation describes the ground state occupancy with a term, $-R_p$, due to pumping and a term, $+\gamma_{10}N_1$, due to decay from the lower lasing level.

Since the three-level laser will have a very fast decay from the lower lasing level to the ground state, it's a good approximation to set $N_1 = 0$. In other words, the lower lasing level is emptied as soon as it gets occupied. In that case, the second equation is solved and the other two equations are decoupled.

$$\frac{\mathrm{d}N_2}{\mathrm{d}t} = R_p - \gamma_{21}N_2 - \sigma_{21}N_2\left(\frac{I}{h\nu}\right) \tag{6.16}$$

$$\frac{\mathrm{d}N_0}{\mathrm{d}t} = -\frac{\mathrm{d}N_2}{\mathrm{d}t}. \tag{6.17}$$

This allows us to solve for the steady-state irradiance. In the steady state, $\frac{\mathrm{d}N_2}{\mathrm{d}t} = 0$, leading to

$$I = h\nu\frac{R_p - \gamma_{21}N_2}{\sigma_{21}N_2}. \tag{6.18}$$

If we want any output at all, $I > 0$, we need to pump faster than the rate at which atoms spontaneously decay back to the ground state: $R_p > \gamma_{21}N_2$. This is just the condition for maintaining a population inversion. We see that as in the two-level system, the output irradiance is simply proportional to the population inversion.

These equations indicate that the optical gain is positive provided a population inversion can be maintained. However, we have not yet incorporated the possibility of loss of photons as they traverse the active medium or interact with the cavity mirrors. The next section addresses that issue and we find that we need to pump a little harder to make up for those optical losses.

6.1.5 Lasing Threshold

Not all photons in the irradiance I have a chance of producing stimulated emission. In real lasers, photons are lost due to scattering in the gain medium, by transmission through the laser cavity mirrors, by scatter or absorption on the cavity mirrors, due to misalignment of the cavity mirrors, and so forth.

Imagine a gain medium with some gain $\frac{dI}{dz} = g$ due to a population inversion maintained by a constant pump rate, as described in the last section. Consider a collection of photons going a whole round trip in the laser cavity, bouncing off each cavity mirror in turn and coming back to their starting point within the gain medium. If the irradiance corresponding to that collection of photons was I_n before the round trip, it will be

$$I_{n+1} = R_1 R_2 (1 - \beta) e^{2gL} I_n, \tag{6.19}$$

where L is the cavity length, and the round trip optical loss excluding loss due to transmission through the mirrors is β. The cavity mirror reflectances are R_1 and R_2 respectively. The exponential is due to the constant gain $\frac{dI}{dz} = g$ within the medium. If the irradiance is to increase due to the gain g, then we need $I_{n+1} > I_n$. This implies

$$g > \left| \frac{1}{2L} \ln [R_1 R_2 (1 - \beta)] \right|. \tag{6.20}$$

Until the round trip gain reaches this threshold, the laser will remain dark. This is the reason the active medium generally needs to be embedded between two well-aligned, high reflective mirrors. If the mirrors are misaligned or damaged then the optical loss, β, is significantly larger than 0. The argument of the logarithm is then significantly smaller than one, requiring higher optical gain to compensate for the loss. Such high optical gain is not usually available.

6.2 Etendue and Radiance

Perhaps the most outstanding property of lasers, after coherence, is their low etendue and correspondingly high radiance. To understand these related properties, we consider light leaving some planar region of area ΔA. Some or all of the light enters an optical system whose aperture spans the solid angle $\Delta \Omega$ and lies in a direction θ with respect to source normal (see Figure 6.7). The source area, ΔA, projected onto a plane perpendicular to the average direction of travel of the light is $\Delta A_\perp = \Delta A \cos \theta$. If the beam emitted from the source doesn't fill the aperture, then we use the solid angular spread of the size of the beam at the aperture instead. The product

$$\Delta G = \Delta A_\perp \Delta \Omega \tag{6.21}$$

is known as the etendue.

Etendue is an important characteristic of an optical system. Etendue conservation can be considered as equivalent to the conservation of rays in the geometric optics picture. Etendue can only decrease if rays are removed from the optical system. (As usual, the

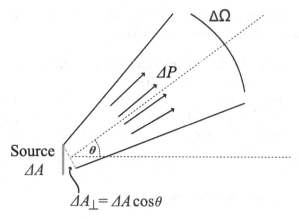

Figure 6.7 Construction for average radiance and etendue. The power entering the solid angle $\Delta\Omega$ is ΔP, which is emitted from the area ΔA. The area of emission ΔA has an apparent size (projection) $\Delta A_\perp$ when viewed from direction θ.

word "rays" is a stand-in for "wavefront normals".) For example, blocking rays at an iris reduces the cross-sectional area of the remaining rays, reducing the etendue. So etendue can only decrease if optical power is also being discarded. On the other hand, if light scattering occurs, such as from a diffuse source, this increases the angular spread of the rays and etendue increases. Cutting out rays seems to make the ray bundle "simpler" while scattering seems to "add more chaos," so it shouldn't be a surprise that etendue is also related to the entropy of the rays. Well-designed optical systems tend to conserve etendue because they don't throw away any light unnecessarily and the optics are good enough that they don't induce scatter or significant random wavefront distortion. Note that ΔG is a function of the angle θ but doesn't include any information about the power radiated into that particular direction. It's a fully geometric quantity. We incorporate the power by defining a new quantity known as the radiance.

Radiance is a measure of the "brightness" of a source. It's the power emitted from a source per unit projected area of the source, per unit solid angular spread of the rays leaving that area in some direction (θ, ϕ).

$$L(\theta, \phi) = \frac{\partial^2 P}{\partial A_\perp \partial \Omega}. \tag{6.22}$$

Consider the source in Figure 6.7 again. The average radiance captured by the optical system covering solid angle $\Delta\Omega$ is

$$\bar{L} = \frac{\Delta P}{\Delta A_\perp \Delta\Omega}. \tag{6.23}$$

The quantity in the denominator is the etendue of the bulb-lens system while the numerator is the power entering the system. Since etendue conservation implies that no power is lost, then average radiance must be conserved by an optical system whenever etendue is conserved. Lasers tend to have the highest radiance of all artificial light sources. Even though the total power is modest, the light is emitted from a very small area into the smallest possible (diffraction limited) solid angle. This is a direct consequence of spatial coherence.

Note that we can write the irradiance as $I = \frac{\Delta P}{\Delta A} = L\Delta\Omega$. Since $\Delta\Omega$ is under our control through the use of lenses, lasers have the highest irradiance of all sources when focused.

Example 6.1 Collimation and Focusing The etendue determines how well a beam can be collimated and how tightly a beam can be focused. Consider a lens of focal length f and area A_{lens} that is fully illuminated by light that is at least approximately collimated. The solid-angular spread of rays coming to a focus downstream of a lens whose focal length is large compared to its diameter is

$$\Delta\Omega \approx \frac{A_{\text{lens}}}{f^2}. \tag{6.24}$$

The etendue of the beam at the focus is then

$$\Delta G = \Delta\Omega\Delta A = \frac{A_{\text{lens}}}{f^2}A_{\text{focus}}. \tag{6.25}$$

Clearly, we get a smaller focus for lower values of etendue. Since ΔG is conserved for the optical system, the etendue before the lens should be made as small as possible in order to get the smallest possible A_{focus}. Lasers have much lower etendue than other sources of light. Therefore, the smallest focus is achieved with a laser. Also, the presence of f^2 in the denominator indicates that reducing the focal length of the lens decreases the size of the focus.

Example 6.2 Etendue Conservation in a Simple Optical System We will consider the etendue of the optical system shown in the figure below. For simplicity, I'll assume the source is something like a small frosted light bulb that we can model as a uniformly radiating sphere with some radius r. (This source is also "Lambertian," which means that it's surface looks equally bright regardless of the orientation of the surface with respect to the viewer.)

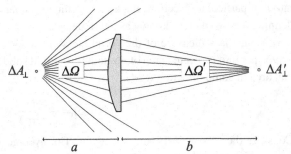

The projected area of the source is then $\Delta A_{\perp} = \pi r^2$. The lens subtends a solid angle $\Delta\Omega = \frac{\pi d^2}{a^2}$ where d is the lens radius and a is the source-lens distance. So, the etendue at the source, of the portion of light entering the optical system, is $\Delta G = \frac{\pi^2 r^2 d^2}{a^2}$. After the lens, the image occurs a distance b after the lens and the magnification is $\frac{b}{a}$. The image area is then $\Delta A'_{\perp} = \pi r'^2 = \pi\left(\frac{rb}{a}\right)^2$ and the solid angle subtended by the beam approaching the image is $\Delta\Omega' = \frac{\pi d^2}{b^2}$. The etendue at the image is then $\Delta G' = \frac{\pi^2 r^2 b^2}{a^2}\frac{d^2}{b^2} = \Delta G$. No change in the etendue occurred due to the lens.

Example 6.3 Radiance of a Light Bulb Versus a HeNe Laser Consider the case of a person looking directly at a typical light bulb which radiates say 5 Watts of power at visible wavelengths, approximately uniformly in all directions. The perpendicular area of a typical filament is say $\Delta A_\perp \approx \pi(0.5 \text{ mm})^2 \approx 0.8 \text{ mm}^2$. If one is considering the light entering a dilated pupil (diameter 5 mm) at a distance of say 1 m, then the solid angle subtended by the pupil is $\Delta \Omega \approx 8 \times 10^{-5}$ sr. The power entering the pupil is $\Delta P = 1 \times 10^{-4}$ W, leading to a radiance estimate of $L \approx 2 \text{ W sr}^{-1} \text{mm}^{-1}$. For comparison, consider staring directly into a 5 mW HeNe laser with beamwidth shown in Figure 6.9. The waist is at the laser aperture and has a radius of about 0.6 mm. The waist has a projected area $\Delta A_\perp \approx \pi(0.6 \text{mm})^2 = 1.1 \text{ mm}^2$. By the time the beam reaches 1 meter from the waist it has a radius of about 0.75 mm and corresponding solid angle divergence of $\Delta \Omega \approx 1.4 \times 10^{-6}$ sr (starting from Eq. (4.15)). The power entering the eye is the full power of the laser, $\Delta P = 5$ mW leading to a radiance estimate of $L = 3000 \text{ W sr}^{-1} \text{mm}^{-1}$. This is 1,500 times higher than the radiance of the filament. We know that it's uncomfortable to stare at a light bulb filament directly. The comparatively enormous radiance of the laser should give pause to any fleeting thoughts of looking into a laser aperture to see if it's turned on!

Exercises

6.1 In the case of atoms or molecules with degenerate energy levels, Eq. (6.1) should be modified so that the right-hand side of each equation is multiplied by an integer representing the degeneracy of the corresponding energy level. The degeneracy of level E_1 is g_1 and the degeneracy of level E_2 is g_2. Carry these coefficients through to Eq. (6.12) to see how the result is affected.

6.2 If N_1 and N_2 are independent of location in the gain medium, Equation (6.12) indicates exponential growth in the flux S as a function of distance. Comment on why this assumption might break down as I increases. How would you expect the behavior to be modified?

6.3 Choose a particular laser type (NdYAG, Ruby, Ti:sapphire, Argon Ion, CO_2, etc.) to read about. Draw an energy level diagram of the levels involved. Annotate the diagram with the level energies in eV and identify the transition involved in the lasing, pumping, and so forth. Comment on the function of each transition.

6.4 Show that Eq. (6.20) follows from Eq. (6.19).

6.5 You test a particular type of laser with a 30 cm long cavity (filled with gain medium) and no optical loss. You find that with perfect alignment of the end mirrors, it will only lase if the mirror reflectivities are higher than 90%. You now replace one of the

90% cavity mirrors with an 80% reflective cavity mirror. How much longer will you have to make the laser cavity if you still want it to lase?

6.6 You're attempting to get a HeNe laser to reach threshold. When the laser cavity is perfectly aligned, it will lase provided $R_1 R_2 (1 - \beta) > 0.95$. The cavity mirrors have $R_1 = 0.99$, $R_2 = 0.98$ and have no intrinsic optical loss. The cavity mirrors are located 20 cm away from the end of the 20 cm capillary that has a diameter of 200 microns. Make a rough estimate of the allowable angular misalignment of a cavity mirror beyond which loss due to beam clipping makes lasing impossible.

6.7 Using Eqs. (6.13–6.15) as a reference, write down a set of three-level rate equations that describe the type of three-level lasers where the lasing transition is $1 \to 0$ rather than $2 \to 1$.

6.8 Approximate the etendue of the human eye. For the purposes of this problem, use the central retina, 6 mm in diameter, as the active area when looking in a single direction. The distance from the lens to the retina is about 20 mm. Take the diameter of the pupil as 4 mm.

6.9 A 10 Watt incandescent light bulb filament is imaged onto a screen 150 cm away using three equally spaced positive lenses of equal focal length, $f_1 = f_2 = f_3 = 12.5$ cm. The lens diameters are $d_1 = 5.0$ cm, $d_2 = 2.5$ cm, and $d_3 = 5.0$ cm, respectively. Treat the filament as an isotropic radiator with surface area of 1 cm^2. Is etendue conserved by the system? What is the minimum etendue of any element (and therefore the overall etendue of the system)?

6.10 Do the last problem but replace the incandescent light bulb filament by a 5 mW laser with a $\omega_0 = 1$ mm waist at the laser aperture. Assume that the lenses form beam waists at the locations of the images in the last problem.

6.11 Calculate the radiance of each source and list in order from highest to lowest: HeNe laser, LED, incandescent light bulb, campfire, and the sun.

6.12 A 10 Watt incandescent light bulb filament is imaged onto a screen 120 cm away using a lens with diameter 5 cm and focal length 20 cm. An image can be formed by placing the lens at either of two distances from the filament. If the filament has radiance $L = 1 \times 10^6$ W/m^2/sr, what is the radiance of the image for each location of the lens? How does this comport with observations of the apparent image brightness in each case? Treat the filament as an isotropic radiator with surface area of 1 cm^2. *Hint*: The image on the screen radiates approximately isotropically into a hemisphere of solid angle 2π sr.

6.3 Experiment: Helium–Neon Laser

Objectives

1 Observe the emission spectrum of an excited mixture of helium and neon gas – the active medium in a HeNe laser.

2 Get the laser to lase by providing sufficient optical gain through good alignment of the laser cavity mirrors.

3 Measure the properties of the output beam and show that it is indeed a diffraction limited Hermite–Gaussian beam.

Equipment Needed

- Helium-Neon laser tube securely mounted with *covered and appropriately protected electrical contacts*. The tube should have windows or Brewster plates on the ends, not mirrors.
- A Helium-Neon laser power supply *that is safe for use in a classroom setting*.
- High-reflective end mirror, $R \geq 99\%$.
- Input mirror, $95\% < R < 99\%$. Polished on both sides.
- CCD or CMOS camera with manual aperture adjustment. A camera with a bare image sensor is ideal. Otherwise a diffuser may be needed. ("Scotch tape"[5] works well.)
- A set of neutral density filters or similar filters to reduce beam irradiance to levels manageable by the camera. (A pair of uncoated glass wedges or prisms can also be used.)
- A visible-light, digital, fiber-coupled spectrometer.

The Excited He-Ne Gas Mixture

Consult with the instructor on safe setup of the HeNe tube. It is very important not to touch the two ends of the He-Ne tube that may be at very high voltage (thousands of Volts). Inadvertently touching the terminals of the laser tube may result in a dangerous electric shock.

Turn the laser tube on. Use a spectrometer pointed at the middle of the He-Ne tube (from the side) and observe the spectrum of the glow from the helium and neon gas mixture. Make a note of the most prominent lines and search for the neon line at 632.8 nm. This is the line on which the gas will lase in the following section. Helium-Neon lasers can also be induced to lase on other lines. Look up their wavelengths and see if you can find those lines. Make a note of the relative strength of the lasing lines that you can find. Is there anything else remarkable about these lines compared to other lines in the discharge glow?

Recommendation: You may need a lens to focus light into the fiber-coupled spectrometer.

Aligning the Laser

Align the open laser cavity so that the laser is resonating in the TEM_{00} mode.

---

[5] "Scotch tape" is a matte, translucent, plastic, adhesive tape.

Recommendations: Make sure you have appropriate mirrors for the cavity. Check with your instructor on their correct placement in order to ensure an optically stable resonator (see Section 7.3.1).

Getting the laser to lase depends on minimizing optical losses and maximizing the gain in the resonant cavity within which the laser tube is located. In other words, you must set up the cavity mirrors (input coupler and end mirror) exceedingly carefully. They must be placed so that the resonant beam will travel precisely along the axis of the laser tube with the "tails" of the beam profile "just scraping" the edge of the capillary along the length of the tube. This requires thoughtful placement of the mirrors in the transverse directions but also requires them to be oriented precisely in pitch and yaw. Because of the precision required, it is common to align the laser with the help of a second laser with a similar beam profile as the one you are expecting to obtain. Another HeNe is ideal. See Figure 6.8 for the first step in the alignment: finding the axis of the laser tube. Start by getting light though the laser tube's capillary without the cavity mirrors in place. Then insert the mirrors one at a time starting with the one on the far side of the tube from the alignment laser. Make sure the reflected beam from each mirror goes straight back up the optics chain and is incident on the alignment laser aperture. If the final alignment is good enough that the light from the external laser flashes in the tube due to self-interference, then its a good bet the laser will be close to lasing when the tube is turned on. You may need to fiddle a little with the pitch and yaw adjustment of the mirrors to search for the resonant condition even after you've done your best to align them. If, after an hour or two of alignment work, you are still unsuccessful, it is time to double check that you have the correct optics, that all optical surfaces are completely clean. If necessary, reevaluate your alignment methodology. It usually takes a couple of hours to get the alignment right the first time.

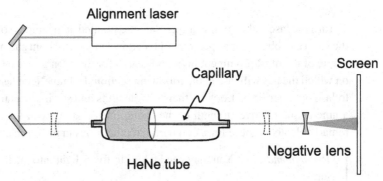

Figure 6.8 Schematic for finding the axis of the laser tube. The cavity mirrors' intended locations are shown with dashed lines. The input mirror will be on the left, on the same side as the alignment laser. The end mirror will be on the right.

Analyzing the Beam Produced

To fully characterize the beam exiting the laser, you will measure the beam radius at several locations downstream of the input mirror. The beam that exits the input mirror of the resonant cavity will generally have a waist somewhere within the laser tube. The waist location and beam radius at the waist are two parameters that uniquely describe the output beam of the laser (or any Gaussian beam). You will fit the irradiance profile at several locations downstream of the input mirror to obtain the beam radius at those locations.

The fit function should be of the form

$$I(x) = I_0 TEM_{00}(x - a, 0, w) + B, \tag{6.26}$$

where $I(x)$ is the value of the matrix element corresponding to the pixel at position x, $TEM_{00}(x, y, w)$ is the expected functional form of the irradiance in the xy-plane perpendicular to the beam, and $y = 0$ corresponds to the center of the beam in the vertical direction. w is the beam radius and the offset a accommodates the fact that the beam is not likely to be perfectly horizontally centered on the image sensor. B represents the signal of the background light or other constant offset affecting the irradiance. Thus, the fit will have four fit parameters: I_0, a, w, and B. The fit will need to be performed using a nonlinear fitting routine since the fit function is not linear in the fit parameters. (See Section 2.3.2.)

The beam radius can also be found by moving a knife-edge across the beam and monitoring the fraction of the total power that is passed by the edge as a function of position (see Exercise 4.5 in Chapter 4). The most accurate and easiest (and least hands-on, most expensive) way of obtaining beam radii is to use a dedicated "beam scanner" tool sold by optical equipment manufacturers. A beam scanner continuously scans an aperture over the beam in order to estimate its dimensions.

After you have obtained the beam radius at several locations downstream of the laser, you will plot the beam radius as a function of position and fit that data to the expected form in order to find the waist location and waist width. An example of a fit similar to the one you should produce is shown in Figure 6.9.

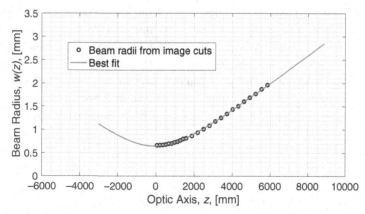

Figure 6.9 Beam radius as a function of distance on the optic axis with a fit to the expected dependence.

Recommendations: The irradiance of the light may vary a lot as the beam size changes. You will need to be very careful not to saturate the image sensor in the camera. For this reason, make sure that the beam is first passed through neutral density filters or other appropriate filters to cut down the irradiance. An alternative way to cut down the irradiance is to reflect the beam from an uncoated, flat, polished glass surface such as the surface of a prism. The reflected beam will have about 4% of the incident power if the reflection is close to normal incidence. You will probably need to reduce the beam irradiance by two or more orders of magnitude in order not to saturate the image sensor. A pair of prisms will cut the beam irradiance down by a factor of $0.04^2 = 1.6 \times 10^{-3}$. It's very common to underestimate the amount of attenuation needed so be sure to use enough. Fringes on the camera can also be a problem, especially for image sensors with glass over them. You can choose to ignore them and deal with them appropriately during the analysis. If you prefer a hardware fix, a diffuser placed directly onto the glass covering the image sensor can be the answer. (Sticking a small piece of "Scotch tape" right onto the image sensor glass works very well for this.) The glass covering will need to be right up against the image sensor for this to work. Even so, the diffuser won't be directly on the sensor, and it may be wise to calibrate against a known beam. During the analysis, I recommend leaving the experiment set up in case you need more data. Given how difficult it can be to get the laser properly aligned, it's best not to take this experiment down until analysis of the data is complete.

Ideas for Further Investigation

Get the laser to resonate on the TEM_{10} or TEM_{01} mode. Characterize the output beam in the same way as you characterized the TEM_{00} earlier. Do you find that the higher order nature of the mode affects the beam propagation ($w(z)$, z_0, waist location,...) in any way? Should it?

It could be interesting to compare two or all three methods discussed earlier for measuring the width of a beam.

For an advanced challenge, you could try getting one of the other lines to resonate, thereby producing a differently colored HeNe: orange at 612 nm, green at 543 nm, and so on. You would need new cavity mirrors tailored to each of these lines. Getting them to lase is also difficult because these lines are weaker than the 633 nm line. It may not be possible to accomplish this with the equipment available to you but it's interesting to consider the challenge. If you're interested, read up on how others have achieved this.[6] If you do achieve resonance on a different line, how do the beam properties compare with those of the 632.8 nm line (power, stability, polarization, etc.)?

---

[6] A longstanding online resource is "SAM's Laser FAQ." Look for the HeNe section.

7   Optical Cavities

7.1   Use of Optical Cavities

Optical cavities form the core of lasers. As we saw in Chapter 6, a gain medium is placed within an optical cavity in order to get enough optical gain to allow lasing to occur. Optical cavities are also used without a gain medium for their resonant properties; such "external" optical cavities form the core of many of the most sensitive instruments in use today. Due to their high sensitivity, optical cavities are now used very widely. Applications include laser frequency stabilization, laser spectroscopy, gravitational wave detection, quantum mechanics experiments, and nonlinear optics, to name but a few.

7.2   Plane-Wave Cavity

The simplest resonant optical cavity consists of two mirrors facing one another. Light enters the cavity through one of the mirrors, which is partially transmissive. The distance between the mirrors is fixed so that the light bouncing back and forth between the mirrors executes an integer number of oscillations per round trip and therefore combines constructively with light just entering the cavity, thereby forming a driven standing electromagnetic wave. Keeping the optical pathlength between the mirrors at the precise length required for resonance is a significant technical challenge that we will ignore. Also, we don't address the fact that a real cavity would have finite mirrors. To keep the beam from spilling out of the side of the cavity, one or both of the mirrors would need to be slightly concave. We simply start with the idealized case of infinite flat mirrors and a beam composed of monochromatic infinite plane waves. Figure 7.1 shows the cavity and the amplitudes of the various fields involved.

The input beam comes from the laser and impinges on the input mirror. The other mirror is usually known as the end mirror. To minimize optical loss due to the mirror substrates, the input mirror and end mirror are oriented so that the reflective coatings are on the cavity-side of the optics. Therefore, the input beam actually reflects from the input mirror coating inside the glass substrate of the mirror. The field approaching the (reflective surface on the) input mirror is a plane wave

$$\vec{E}_i = \hat{n}\, u_i e^{-i(kz - \omega t)}, \tag{7.1}$$

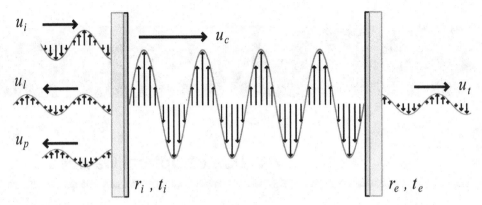

Figure 7.1 A plane-wave cavity. The reflective optical coatings are normally on the cavity side of the mirror substrates as shown.

where $\hat{n}$ is the polarization. The complex field amplitude u_i is a *real* constant since we also choose the overall phase offset to be zero at the input mirror. (Since we are treating beam as a plane wave, u_i is not a function of position.) The wave number and angular frequency of the plane wave are k and ω respectively. If we take $z = 0$ at the input mirror

$$\vec{E}_i = \hat{n} u_i e^{i\omega t}. \tag{7.2}$$

Immediately to the right of the input mirror, the cavity field is

$$\vec{E}_c = \hat{n} u_c e^{i\omega t}. \tag{7.3}$$

Since cavity traversal takes an integer number of oscillations, the phase after any traversal is the same as that of the input field, $\vec{E}_i$. To find the superposition of any fields present at the input mirror then, we just need to add their (signed) amplitudes.

The direct reflection of the input beam from the input mirror coating is u_p. This is known as the "promptly" reflected beam to distinguish it from that part of the beam that enters the cavity and then leaks back out the input mirror. That "leakage beam" with amplitude u_l combines with the promptly reflected beam u_p to make the total reflected beam, $u_r = u_p + u_l$. The amplitude reflection coefficients of the input mirror and end mirror are r_i and r_e, respectively. (The amplitude reflection coefficient is the ratio of the reflected field amplitude to the incident field amplitude.) Similarly, t_i and t_e are the amplitude transmission coefficients. For the time being, we assume that the end mirror is perfectly reflective, $|r_e| = 1$, while the input mirror is very slightly transmissive, $|r_i| < 1$. That will allow light to enter the cavity through the input mirror.

As stated earlier, light will form a standing wave in the cavity when the length of the cavity is such that one round trip from the input mirror to the end mirror and back again accumulates an integer number of 2π radians of phase. But how does any significant field build up in the cavity when the input mirror is only very slightly transmissive? After all, most of the input light would seem to be reflected back toward the laser. The explanation lies in the fact that the promptly reflected beam is 180° out of phase with the leakage beam, so u_p and u_l have opposite sign. They interfere destructively, meaning that no power can leave the cavity through the input mirror until $|u_l| \geq |u_p|$.

Imagine, for example, that the cavity is at the correct length to be resonant and the input beam is suddenly turned on. Although only a small fraction of the input beam's light enters the cavity through the input mirror, the cavity field nevertheless starts to build up. Equilibrium is reached when the leakage beam's amplitude is sufficiently larger than the promptly reflected beam's amplitude that the power leaving the cavity matches the power entering. If the input mirror and end mirror are highly reflective and there are no significant losses in the cavity, the cavity field can become quite large.

To find the steady-state relationships between the various fields, we note that once the power flowing into the cavity equals the power leaving the cavity, equilibrium has been achieved. Since we are dealing with plane waves, the irradiance is uniform. The irradiance of the input beam as it approaches the cavity is

$$I_i = \frac{1}{2}\epsilon_0 c u_i^2, \tag{7.4}$$

where ϵ_0 is the permittivity of free space and c is the speed of light. Assuming no cavity losses and $|r_e| = 1$, the total power per unit area leaving the cavity is

$$
\begin{aligned}
I_{\text{out}} &= \frac{1}{2}\epsilon_0 c u_r^2 \\
&= \frac{1}{2}\epsilon_0 c \left(u_p + u_l\right)^2 \\
&= \frac{1}{2}\epsilon_0 c \left(r_i u_i - t_i u_c\right)^2 .
\end{aligned}
\tag{7.5}
$$

The relative minus sign between the terms in the last line is due to the 180° phase shift of the prompt beam relative to the leakage beam. Energy balance is achieved when

$$I_i = I_{\text{out}}. \tag{7.6}$$

This gives (see Exercise 7.1)

$$\frac{u_c}{u_i} = \frac{r_i}{t_i} + \sqrt{\left(\frac{r_i}{t_i}\right)^2 + 1} \tag{7.7}$$

$$= \frac{r_i + 1}{t_i}. \tag{7.8}$$

As a check, we take the limit in which the input mirror to the cavity and becomes perfectly transmissive: $r_i \to 0$ and $t_i \to 1$. The cavity amplitude just becomes equal to the input amplitude, as expected. On the other hand, if the input mirror is highly reflective, $t_i \ll 1$ then the ratio of the cavity field amplitude to the input field amplitude is

$$\frac{u_c}{u_i} \approx \frac{2}{t_i}. \tag{7.9}$$

If one also uses an end mirror that has nonzero transmission the equivalent expression is

$$\frac{u_c}{u_i} = \frac{1}{\eta}\left[\frac{r_i}{t_i} + \sqrt{\left(\frac{r_i}{t_i}\right)^2 + \eta}\right], \tag{7.10}$$

where $\eta = 1 + t_e^2/t_i^2$. The square of this quantity is called the intracavity power buildup. It's the ratio of the power circulating in the cavity to the power incident on the cavity

$$\frac{P_c}{P_i} = \left(\frac{u_c}{u_i}\right)^2.$$ (7.11)

Example 7.1 Cavity with High-Reflectivity Mirrors Consider a 5 mW HeNe laser illuminating a cavity with a 99.5% reflective ($r_i^2 = 0.995$) input mirror and a 99.99% reflective end mirror ($r_e^2 = 0.9999$). The ratio of the power circulating in the cavity to the input power is

$$\frac{P_c}{P_i} = \left(\frac{u_c}{u_i}\right)^2$$

$$\approx 4 * \frac{r_i^2}{t_i^2}$$

$$= 796.$$

This leads to a circulating power in the cavity of 796 × 5 mW, or almost 4 Watts! The resulting beam will be very visible on the cavity mirrors since small coating imperfections will scatter some of this resonating power. By increasing the reflectivity of the input mirror further, large cavity powers can be built up. This allows us to observe the effects of large optical fields E and B without needing enormous lasers, especially if the cavity has a narrow beam waist (Section 7.3).

Derived Cavity Parameters

For convenience, we define a few additional parameters that depend on the mirror reflectivities, cavity length, cavity losses, and so forth. Although these parameters don't contain any new information per se, they are still useful and extensively quoted.

Free Spectral Range, *FSR*: One of the things we can imagine doing is to change the length of the cavity slightly, by moving one of the cavity mirrors. The cavity resonates when it contains an integer number of half wavelengths, so each time an additional half-wavelength fits into the cavity, a new resonance is encountered. Similarly, if we leave the cavity length fixed but change the wavelength (and therefore frequency) of the laser, the cavity will resonate every time an integer number of half-wavelengths fits within the cavity. If we express the frequency of the laser when the cavity resonates as f_n, then

$$L = \frac{n\lambda_n}{2} = \frac{nc}{2f_n}$$

$$\Rightarrow f_n = \frac{nc}{2L}.$$ (7.12)

Figure 7.2 shows a model for how the power in the cavity varies with laser frequency. The frequency difference between adjacent resonances is known as the "free spectral range"

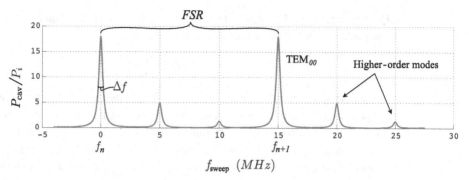

Figure 7.2 The power in a 10 cm symmetric cavity illuminated with a 1,064 nm laser as a function of small changes in the laser frequency. This cavity has a $g_1 = g_2 = 1/2$.

or *FSR*. It is typically a very small fraction of the laser frequency itself. The free spectral range is given by

$$
\begin{aligned}
FSR &= f_{n+1} - f_n \\
&= \frac{c}{2L}\left[(n+1) - n\right] \\
&= \frac{c}{2L}.
\end{aligned}
\tag{7.13}
$$

Cavity Round Trip Loss, β: So far, we haven't considered cavity losses except that cavity losses affect the performance of the cavity exactly like leakage out through the end mirror. (The leakage field leaving the cavity through the input mirror is an active participant in the resonance condition, so transmission of the input mirror affects the resonance differently than transmission through the end mirror or other optical losses.) Typical causes of loss are misalignment, scatter, absorption, polarization rotation, and depolarization. In this section, we will lump all forms of loss into one number, the "cavity round trip loss," β. It's the fractional power lost to the cavity beam during a single round trip, other than by leakage through the input mirror or end mirror.

Finesse, $\mathcal{F}$: This unitless quantity is related to the number of round trips that a photon can be expected to make before dropping out of the cavity, either through scatter, absorption (or other loss mechanisms) or by leaving through the input mirror. It is defined as

$$
\mathcal{F} = \frac{FSR}{\Delta f},
\tag{7.14}
$$

where Δf is the width of the resonance peaks as shown in Figure 7.2. Δf, is defined as the full width of the resonance peak halfway down from it's maximum value – often referred to as the "full width, half max width" or FWHM. High-finesse cavities ($\mathcal{F} \gtrsim 1,000$) trap light in the cavity for longer and have a lot more power buildup compared to low-finesse cavities.

Quality factor, Q: The quality factor can be defined for any resonator. It is the ratio of the resonance width, Δf, to the resonance frequency, f.

$$Q = \frac{f}{\Delta f}. \tag{7.15}$$

In the case of an optical resonator as discussed here, the frequency f of the resonance is just the laser frequency. If we were to take an optical cavity that is resonating and suddenly turn off the input laser light, the cavity light would quickly decay due to light leaving through the input mirror and through cavity losses. The shape of the decay, as for any underdamped harmonic oscillator, is a decaying exponential with time constant τ_e.[1] In terms of the cavity Q, the time constant can be written

$$\tau_e = \frac{Q}{\pi f} = \frac{Q\lambda}{\pi c}, \tag{7.16}$$

where λ is the laser frequency and c is the speed of light, as before. The finesse and the quality factor are related by

$$Q = \frac{\mathcal{F}c}{\lambda (FSR)}. \tag{7.17}$$

Cavity buildup P_{cav}/P_i: The cavity buildup, is the ratio of the power in the cavity on resonance, P_{cav} to the input light power P_i. For resonators with $Q >> 1$, the quality factor can be expressed as the fractional oscillator energy lost per cycle: $Q = 2\pi \frac{E}{\Delta E}$. We can use this fact to obtain the general relationship between the cavity buildup and finesse. (See Exercise 7.7 for the lossless cavity version.)

$$\frac{P_{\text{cav}}}{P_i} = \frac{1}{\pi} \mathcal{F}. \tag{7.18}$$

Comparing this equation with Eq. (7.9) shows that for a lossless cavity with zero end mirror transmission, $\mathcal{F} = \frac{4\pi}{T_i}$. Here, $T_i = t_i^2 \ll 1$ is transmissivity of the input mirror (ratio of transmitted power to incident power). When cavity losses are included, the result is

$$\mathcal{F} = \frac{4\pi T_i}{(T_i + T_e + \beta)^2}, \tag{7.19}$$

where β is the round trip fractional power loss due to all loss mechanisms except leakage through the input and end mirrors. This equation follows from Eq. (7.18) and Eq. (7.10) under the approximations $T_e + \beta \ll T_i \ll 1$. (See Exercise 7.8.)

7.3 Resonant Modes of a Cavity

Infinite mirrors are an idealization that can't be realized in real optical cavities. The plane-wave analysis serves to illuminate the essential aspects of resonance and gives the correct

---

[1] The time constant is the time it takes the oscillator amplitude to fall to $\frac{1}{e}$ of it's initial value.

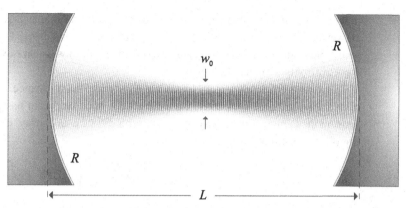

Figure 7.3 A Gaussian beam resonating in an optical cavity. The mirrors are concave to counter the beam spread due to diffraction. At the mirrors, the curvatures of the wavefronts match those of the mirrors. In order to show this, we chose the fairly long wavelength of a CO_2 laser, 10.6 μm, and a short cavity, about 1 mm. In the beam, darker shades of grey indicate higher field magnitudes. The white curves within the beam are the locations where the standing wave has nodes ($\vec{E} = \vec{B} = 0$). The nodes are parallel to the wave fronts with two nodes per wavelength.

field amplitude ratios under idealized conditions. However, it doesn't address the issue of diffraction in beams of finite width and the consequent need to confine the beam in the transverse direction. We use curved mirrors to counteract the propensity of beams to spread out due to diffraction. As in the plane wave case, a standing wave can occur when a round trip in the cavity accumulates an integer number of 2π radians. In addition, the beam must be replicated precisely after each round trip traversal of the cavity. In other words, the beam must overlap with itself after each round trip. In Section 4.1.1 we saw that TEM_{mn} beams are the only beams that retain their shape during propagation. Therefore the resonant modes of optical cavities must be TEM_{mn} modes or linear combinations of TEM_{mn} modes. Provided the mirror radii of curvature and their positions are such that the beam retraces its path on each cavity traversal *and* accumulates an integer number of 2π radians in the process, the beam will form a standing wave in the cavity. A diagram of this is shown in Figure 7.3.

7.3.1 Cavity Stability

A beam that encounters a mirror will reflect back upon itself and retrace its path, *provided* the wave front is parallel to the mirror surface everywhere. Consider for simplicity, a symmetric cavity (identical mirror curvatures) then the beam waist will be in the center of the cavity. At the waist, the wavefronts are flat but as the beam propagates from the waist to either cavity mirror, the wavefronts become more curved due to diffraction. If on reflection, the wavefronts' curvature matches that of the mirrors, then the beam should be self-replicating and the cavity should be able to resonate. The rate at which the wavefronts curve depends on the size of the waist. For a specific set of mirror curvatures and cavity lengths a beam with a specific waist size may be found that will resonate in the cavity. However, for some combinations of mirror curvature R and cavity length L, no physically realizable beam can resonate because no (positive, finite) waist size exists that would allow the wavefronts to match the mirrors.

This discussion allows us to formulate a criterion for resonance, usually called a "stability criterion," in a symmetric cavity. Namely, once the beam has propagated from the waist at the center of the cavity to the mirror a distance $L/2$ away, its curvature $R(z)$ from Eq. (4.7a) must match the mirror curvature R.

$$R = \frac{L}{2}\left[1 + \left(\frac{z_0}{L/2}\right)^2\right]. \tag{7.20}$$

The possible waist size is limited to $0 < w_0 < \infty$, which is equivalent to $0 < z_0 < \infty$. Since L is finite and nonzero

$$0 < w_0 < \infty \tag{7.21a}$$

$$\Leftrightarrow \quad 0 < z_0 < \infty$$

$$\Leftrightarrow \quad 0 < \frac{2R}{L} - 1 < \infty$$

$$\Leftrightarrow \quad \boxed{\frac{L}{2} < R < \infty} \tag{7.21b}$$

So, symmetric cavities with flat mirrors ($R \rightarrow \infty$) can't resonate, nor can the mirrors be so concave that their radius of curvature is less than half the cavity length. A cavity that can't resonate because no self-replicating beam is possible is called "unstable." Cavities for which one of the inequalities above is close to being violated, will have extremely large beams at the cavity mirrors and the location of these beams will be highly sensitive to slight alignment changes. Light will typically spill off the edges of the mirrors of such cavities.

The stability criterion in Eq. (7.21b) is traditionally expressed in terms of the "g-factor" of the mirrors

$$g \equiv \left(1 - \frac{L}{R}\right). \tag{7.22}$$

In terms of g the stability criterion becomes

$$\frac{L}{2} < R < \infty \tag{7.23a}$$

$$\Leftrightarrow \quad -1 < g < 1 \tag{7.23b}$$

$$\Leftrightarrow \quad 0 < g^2 < 1. \tag{7.23c}$$

The left hand inequality in the last line is always satisfied for symmetric cavities but is included because we want to generalize this result to cavities that are not symmetric. In a cavity where the two mirrors have different radii of curvature, R_1 and R_2, there are two corresponding g-factors: $g_1 = (1 - L/R_1)$ and $g_2 = (1 - L/R_2)$. The stability criterion for such a cavity is

$$0 < g_1 g_2 < 1. \tag{7.24}$$

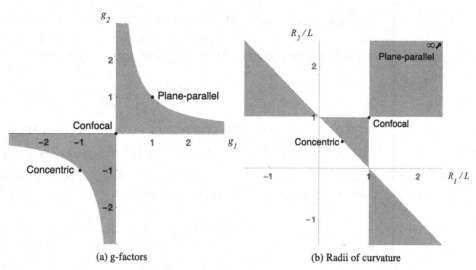

(a) g-factors   (b) Radii of curvature

Figure 7.4 Two ways of visualizing the stability of an optical cavity. Also shown are points corresponding to three marginally stable symmetric cavities: the concentric cavity, the confocal cavity, and the plane-parallel cavity. (a) Shows the stable regions in gray as a function of the mirror g-factors. This is the traditional way of illustrating stability. (b) Shows the stable regions in gray as a function of $\frac{R_1}{L}$ and $\frac{R_2}{L}$. In this view, there is a useful separation of cavity types into separate regions. The central triangle corresponds to cavities with positive sub-confocal (very concave) mirrors. The square region corresponds to cavities with positive super-confocal, (less concave) mirrors. The two wedges on the upper left and lower right extending to infinity correspond to cavities with one convex mirror and one concave mirror.

This relationship can be illustrated on a graph of g_1 vs. g_2 as shown in Figure 7.4. The shaded regions satisfy inequalities 7.24 indicating stable cavities.

Example 7.2 Diffraction Losses   When finite-sized mirrors are illuminated by Gaussian beams, some light always falls off the edge because the Gaussian tails are in principle infinite. In the most common situation, where the beam radius is a couple of millimeters or less and the optics have radii greater than a centimeter, the lost power isn't significant and we ignore it. However, in nearly unstable cavities or in long but stable cavities, the beam radius can get quite large. In those cases, diffraction losses may need to be taken into account. Consider the following sample problem.

Problem. Suppose you're building a stable, $g = g_1 = g_2 = \frac{1}{2}$, $L = 10$ meter long, symmetric optical cavity and using a Nd:YAG, $\lambda = 1,064$ nm laser. How big must the mirrors be in order that diffraction losses are less than 100 parts per million? (100 ppm = 10^{-4}.)

Solution (outline only). We want to find out how much light "falls off the edge of the mirrors," so we will need to find the beam radius w at the cavity mirrors for this particular g. In Exercise 7.11, you find the cavity waist w_0 in terms of R and L. That expression is easy to convert to a function of g. We then use Eq. (4.7b) to get the beam radius w on the end mirrors. After an exciting algebra safari, we find

$$w = \sqrt{\frac{\lambda L}{\pi}} \left(\frac{1}{\sqrt{1-g^2}} \right)^{\frac{1}{2}}. \tag{7.25}$$

So, as $g \to \pm 1$ (plane-parallel or concentric cavity), $w \to \infty$ as expected. For $g = 0.5$ however, we get $w = \sqrt{\frac{\lambda L}{\pi} \left(\frac{2}{\sqrt{3}} \right)} = 2.0$ mm. If we choose optics with a radius r, then we can expect that any light that is outside the optic will be lost. The fraction $f(r)$ of power in a Gaussian (TEM$_{00}$) beam of beam radius w outside a radius r is

$$f(r) = \frac{\int_0^{2\pi} d\theta' \int_r^\infty I(r') r' dr'}{\int_0^{2\pi} d\theta' \int_0^\infty I(r') r' dr'} \tag{7.26}$$

$$= \frac{\int_r^\infty r' e^{-\frac{2r'^2}{w^2}} dr'}{\int_0^\infty r' e^{-\frac{2r'^2}{w^2}} dr'} = e^{-2\frac{r^2}{w^2}}. \tag{7.27}$$

So we need to solve $f(r) < 10^{-4}$ for r with $w = 2$ mm. The result is $r > 4.3$ mm. One would want to provide a margin of safety to account for possible beam motion on the optics and imperfect alignment. Common, one-inch diameter optics, $r = 12.7$ mm, would be more than adequate and even half-inch optics would suffice.

7.3.2 Spatial Modes

The TEM$_{mn}$ modes of the cavity are also known as "spatial modes." Figure 7.5 shows the cross-sectional profiles of the sixteen lowest order spatial modes.

So far we've only discussed the spatial shape of the beams but in order to build up any appreciable power in the cavity, we must actually form a standing wave. In other words, the appropriate TEM$_{mn}$ mode will resonate in when:

1. One round trip in the cavity accumulates a phase of $2n\pi$, where $n = 1, 2, 3 \dots$
2. There is a nonzero amount of power coupled into the resonant mode from some external source.

The first condition means that the precise length of the cavity must be controlled to maintain resonance – a significant challenge. Also, since the higher order spatial modes accumulate more Gouy phase they will resonate at *slightly shorter* cavity lengths than the fundamental (TEM$_{00}$) mode. A demonstration of the dependence of resonance on precise cavity length involves moving one of the mirrors of a cavity using a piezoelectric transducer (a "PZT").[2] While moving the mirror, we record the cavity leakage beam power

---

[2] A PZT responds to electric potential by changing its length slightly in proportion to the potential. By mounting a mirror on such a device, we get a precise position adjustment. Maximum mirror motion is typically on the order of microns which is perfect for moving mirrors distances on the order of a few wavelengths or less in visible or near-infrared light.

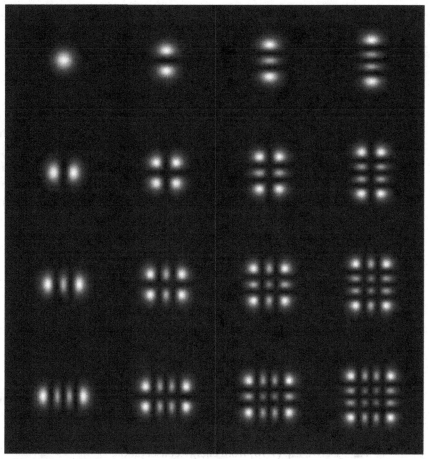

Figure 7.5   The irradiance profiles of TEM$_{mn}$ modes (spatial modes) shown in cross section. The top left mode is TEM$_{00}$, m increases to the right, n increases downward. These irradiance profiles are given by Eq. (4.14).

with a photodiode. A display of the photodiode signal on an oscilloscope shows a peak for each length supporting a resonance. If the mirror moves more than half a wavelength, multiple TEM$_{00}$ peaks will appear, separated by PZT distance changes of $\lambda/2$. Peaks corresponding to the higher order modes will be interspersed with these TEM$_{00}$ peaks. The effect is essentially the same as that shown in Figure 7.2. Since it's the length of the cavity in wavelengths that matters, the cavity responds in the same way whether we decrease the wavelength (increase the frequency) or increase the cavity length.

The Gouy phase accumulated by the modes depends on the Rayleigh range, so the spacing of these mode peaks depends on the g-factors of the cavity. We can even calculate the value of $g_1 g_2$ based on the measured spacing of the modes. (See Exercise 7.10.)

$$g_1 g_2 = \cos^2(\pi\alpha). \tag{7.28}$$

Here, α is the fraction of $\lambda/2$ that the PZT needs to move to go between two adjacent spatial modes. In practice, α is easily measured as the distance between two adjacent spatial mode peaks on the oscilloscope divided by the distance between two TEM_{00} mode peaks. In a properly aligned cavity, the TEM_{00} mode peaks are the tallest peaks and so easily identifiable.

The second boxed condition above reminds us that in order for a mode to resonate, it must be driven by an external source of energy. This could be an active medium as in the case of laser cavity. Here however we are only considering driving the cavity with an *external* beam impinging on the cavity at the input mirror. It's usually desirable to get as much light as possible into the resonant mode. In this case, the cavity should be fed with a beam of the same shape as the beam that will resonate naturally. Recall that the coupling into the cavity relies on destructive interference between the promptly reflected beam and the cavity leakage beam. Therefore, the input beam should ideally be in the same TEM_{mn} mode as the desired resonance and, crucially, have the same geometric parameters: the same waist position and waist size as the resonant beam. Achieving this can be a little tricky; the process is known as "mode matching" and is described in more detail in the following experiment.

Exercises

7.1 Obtain Eq. (7.8) from Eq. (7.6). *Hint*: Taking the plus sign in the quadratic formula leads to $u_c \to u_i$ as $r_1 \to 0$. Taking the negative sign leads to $u_c \to -u_i$ as $r_1 \to 0$. Since the input field and the cavity field are in phase at the input mirror, the plus sign solution is the correct sign to choose.

7.2 Derive Eq. (7.10) from energy conservation.

7.3 (Computer problem) For a lossless plane-wave cavity, plot the cavity power as a function of the input mirror transmission coefficient $0 < t_i < 1$. The transmission coefficient of the end mirror is $t_e = 0.1$. At what value of t_i does the maximum cavity power occur?

7.4 Consider a plane-wave cavity. In Exercise 7.3, you showed that given some value of t_e, there is an optimum value $t_i = t_{\text{crit}}$ that maximizes the cavity power. It turns out that optical losses in the cavity increase t_{crit}. A cavity where $t_i < t_{\text{crit}}$ is known as undercoupled. Cavities with $t_i > t_{\text{crit}}$ are known as overcoupled. A cavity where $t_i = t_{\text{crit}}$ is known as critically coupled. Incorporate optical losses into the analysis given in this chapter and find an expression for t_{crit} in terms of t_e and the round-trip optical loss β. (Here, β is the fractional power lost per round trip not including loss due to transmission through either the input mirror or the end mirror.)

7.5 Show that the free spectral range is a small fraction, $\frac{\lambda}{2L}$, of the laser frequency and that this fraction is only small for cavities of "macroscopic" length.

7.6 How is the energy E stored in a cavity of length L related to the power P_{cav} circulating in the cavity?

7.7 Show that for a lossless cavity, $Q = 2\pi \frac{E}{\Delta E}$ implies Eq. (7.18). *Hint*: First show that for a lossless cavity, ΔE is $\frac{P_{cav}\lambda_n}{c}$. Argue that when the input field is turned off, the energy lost per cycle is four times the energy input by the laser on resonance. Then use the result of Exercise 7.6.

7.8 Show that under the approximation $\beta + T_e \ll T_i \ll 1$, Eq. (7.19) follows from Eq. (7.18) and Eq. (7.10), by replacing $T_e \equiv t_e^2$ with $T_e + \beta$.

7.9 The best mirrors available have optical losses in the parts per million range. What is the maximum finesse cavity you can build with two 10 ppm loss mirrors? (For this problem, assume you can obtain mirrors with any transmissivity you like.) Now assume your symmetric cavity has a 1 cm beam radius on the mirrors. What would the radius of the mirrors need to be in order that diffraction losses are equal to the other losses? Assume the resonance is a TEM_{00} mode and $\beta = 1$ ppm.

7.10 Use equation 4.10 to prove Eq. (7.28) in the case of a semi-symmetric cavity (one flat mirror, $g_1 = 1$, one concave mirror, $0 < g_2 < 1$).

7.11 In any stable resonant optical cavity of length L, the radius of curvature of the wavefront at the cavity mirrors matches the mirrors, R. Using only this information and the propagation characteristics of a Gaussian beam (TEM_{00}) find the beam radius w_0 at the waist of a stable, symmetric cavity. The figure below shows the configuration. (The beam outlines shown are the $\frac{u_0}{e}$ amplitude contours, where u_0 is the field amplitude on the optic axis. Since w_0 is the transverse *radius* at which the amplitude has fallen to $\frac{1}{e}$ of its maximum, the beam *diameter* is $2w_0$.)

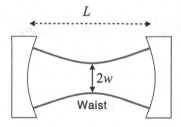

Following the solution outlined in Example 7.2, complete the "algebra safari" to verify Eq. (7.25).

7.12 (Computer problem) Plot the irradiance of the sum of the TEM_{10} and TEM_{01} Hermite-Gaussian cavity modes. This is sometimes called the "donut mode." *Hint*: Use Eq. (4.13).

7.4 Experiment: Resonant Optical Cavity

Objectives
1 Design the optical chain required to obtain a mode-matched optical cavity.
2 Align the cavity and compare the observed cavity output with Hermite-Gaussian modes.
3 Drive the cavity length with a piezo-mounted mirror. Compare the mode-spacing with expectations.

Equipment Needed
- You can use the open-cavity laser you aligned in the helium-neon laser experiment, Chapter 6, or any good quality helium-neon laser (<5 mW).
- Input mirror ($95\% \leq R \leq 99\%$) reflective, flat and polished on both sides.
- End mirror ($99\% \leq R \leq 99.9\%$) radius of curvature ~0.5 m, polished on both sides. Mirror should be mounted on a PZT to allow cavity length adjustments. The PZT should be annular so that light can pass unimpeded through the center of the mirror. If an annular PZT is not available, a folded cavity can be made using an additional, flat, high-reflectivity mirror attached to a PZT or other actuator.
- Function generator capable of driving the PZT with sawtooth or triangle waves in the range 0.1–10 Hz, generating a displacement amplitude of at least $\lambda/2$. Depending on the function generator and PZT, you may also need an intermediate voltage amplifier to provide a high enough voltage to the PZT.
- Access to a set of lenses with a range of focal lengths. A set with about 10–20 lenses and focal lengths between 25 mm and 1,000 mm should be sufficient.
- CCD or CMOS camera with manual aperture adjustment and the ability to take video with at least 15 frames per second.
- A set of neutral density filters or glass wedges to reduce beam irradiance to levels manageable by the camera.
- An optical isolator (e.g. a Faraday isolator).

Driving the Resonant Cavity Modes

Set up a resonant cavity with one curved end mirror and one flat input mirror. Figure 7.6 shows one possible layout. The cavity needs to be optically stable ($0 < g_1 g_2 < 1$). Use the beam from a helium-neon laser to excite optical modes of the cavity. The cavity is not yet mode matched so you may see many higher-order modes being excited.

> *Recommendations*: Place the optical isolator between the laser and the cavity to prevent the output mirror of the HeNe from forming a parasitic interferometer with the cavity you've set up.

Mode-Matching

The goal of mode-matching is to get the incident beam to have the same shape as the beam that resonates naturally in the cavity. This maximizes the power coupled into the

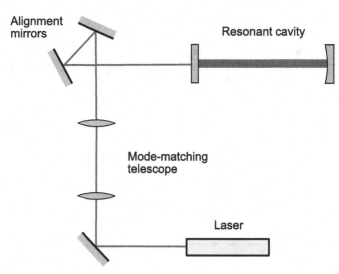

Figure 7.6 A possible layout of the experiment. A screen, camera, or a photodiode is placed behind the end mirror in order to see the resonant mode shapes or monitor the optical power.

cavity. Choosing the correct lenses for the mode-matching telescope and determining their placement can seem daunting. Start by choosing the cavity mirrors and an optically stable cavity length. Use a flat input mirror and concave end mirror. That way, you don't need to compensate for the fact that the input mirror acts as a lens for the beam entering the cavity. Since your input mirror is flat, the cavity waist will be at the input mirror. (The cavity is equivalent to a symmetric cavity of twice the length. In a symmetric cavity, the waist is in the center.) Calculate the waist radius w_c for the beam that naturally resonates in this cavity. (See Exercise 7.11.)

Now make enough measurements of the beam exiting your laser so that you can find the radius of the beam waist w and its location (which may be inside the laser itself). Your job is now to mode match the laser to the cavity – to convert the beam exiting the laser into a beam with a waist right at the input mirror to the cavity and having the exact radius w_c that you found earlier. There are two reasonably simple approaches to mode-matching discussed below. The first method is easier but can take up a fair amount of space. The second method provides an alternative when space is very limited. Whichever method you use, check your solution by propagating (using the ABCD method) the beam from the laser to the cavity through your proposed mode-matching telescope.

Mode-Matching Method 1 – A Collimating Telescope

This approach is the simplest because it involves very little calculation to get a system that is reasonably well mode-matched. The method is based on the observation that the size of a waist is inversely proportional to the steepness (in the far field) with which the beam approaches the focus.

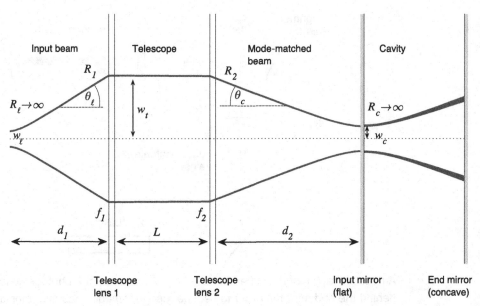

Input beam   Telescope   Mode-matched   Cavity
                           beam

Figure 7.7 A mode-matching telescope with a collimated beam between two lenses. The beam approaching telescope lens 1 diverges with an angle θ_ℓ with respect to the optic axis. The beam leaving telescope lens 2 converges with an angle θ_c with respect to the optic axis.

Consider a telescope like the one shown in Figure 7.7. The width of the collimated beam in the telescope is $w_t \approx \theta_\ell f_1 \approx \theta_c f_2$. Applying Eq. (4.15) gives

$$\frac{w_c}{w_\ell} = \frac{f_2}{f_1}. \tag{7.29}$$

Thus, if one knows the size of the waist in the laser w_ℓ and the target waist size for the cavity w_c, then it's a simple matter to mode-match provided one can find a set of lenses satisfying Eq. (7.29). A significant advantage of this approach is that the distance between the lenses f_1 and f_2 can be allowed to vary quite a bit to accommodate a convenient position for the cavity relative to the laser. Also, if the ratio f_2/f_1 isn't quite right, the distance between these lenses can be varied, providing some "fine adjust" to improve the mode-matching.

The main disadvantage of this approach is the fact that it relies on the lenses being far from the waists $f_1 > z_\ell$ and $f_2 > z_c$ (where z_ℓ and z_c are the Rayleigh ranges, Eq. (4.6), about the laser and cavity waists respectively). However, many lasers have quite long Rayleigh ranges and so f_1 may need to be impractically large. One way around this last problem is to add a third lens before f_1 with a moderately short focal length, perhaps a few hundred millimeters. The new lens brings the beam to a waist that is smaller than the original laser waist. This new waist becomes the new w_ℓ and has a shorter Rayleigh range than the laser. Thus, f_1 can have some reasonably small value. A judicious choice of focal length and position for this new lens can also make it easier to find a pair of lenses satisfying Eq. (7.29).

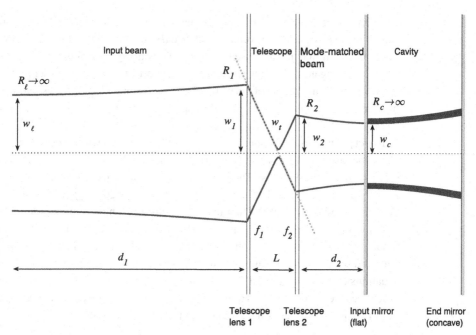

Figure 7.8 A Newtonian mode-matching telescope followed by a cavity. The vertical scale is greatly exaggerated to illustrate the beam profile. Various quantities used in the mode-matching calculation are indicated. The fact that the dotted diagonal line follows the beam width on both sides of the focus, shows that a purely geometric optics description of the telescope is adequate.

Mode-Matching Method 2 – A Newtonian Telescope

This approach is based on the observation that a reasonably short two-mirror telescope where the beam comes to a focus between the lenses is quite accurately described by geometric optics. The main advantage of this approach is that it allows you to use a two lens mode-matching telescope occupying a small space. Finding the right solution is a bit more difficult because now there are three parameters that strongly affect the telescope's operation: f_1, f_2, and also the telescope length L. Referring to Figure 7.8, the equations we use are

$$\frac{1}{f_1} = \frac{1}{R_1} + \frac{1 + w_2/w_1}{L}, \tag{7.30}$$

$$\frac{1}{f_2} = \frac{1}{R_2} + \frac{1 + w_1/w_2}{L}. \tag{7.31}$$

These formulae are derived under the assumption that the telescope is short enough that geometric optics provides an adequate description of its behavior. This requires that the Rayleigh range of the telescope waist be much shorter than the focal lengths of the two lenses: $f_1, f_2 \gg z_t$.

To use this method, first choose some convenient distance d_1 from the laser waist to the first telescope lens (see Figure 7.8). Propagate the beam from the waist w_ℓ to the first telescope lens and obtain the beam size w_1 and radius of curvature R_1 at the first telescope lens. Then choose some convenient distance d_2 between the second telescope lens and the cavity input. (It may make sense to start by choosing d_2 to be about the same as your cavity length.) Propagate the target waist w_c backward to the second telescope lens to find the beam width w_2 and radius of curvature R_2 at the second telescope lens. Now there are three free parameters: the length of the telescope L and the focal lengths f_1, f_2 of the telescope lenses.

To obtain an initial estimate for the focal lengths of the two lenses in the mode-matching telescope, use Eqs. (7.30) and (7.31). Plot f_1 and f_2 as functions of L for some reasonable range of telescope lengths. Then choose the L that gives focal lengths closest to lenses actually available. Note that changing L (without changing d_2) corresponds to moving all the optics downstream of the first telescope lens as a single unit; so you will want to have a solution before you start setting up the telescope. If no acceptable focal lengths can be found, try adjusting d_1 or d_2 and checking again. If very few lenses are available, it may make sense to choose f_1 and f_2 first and then rewrite w_2 and R_2 in the above-mentioned equations in terms of w_c and d_2. That allows you to adjust L and d_2 to accommodate the exact focal lengths chosen. However, the algebra is more complicated and obtaining a useful solution requires some foresight in the choice of f_1 and f_2.

Analyzing the Modes

Once you have a mode-matching solution, spend time carefully aligning the mode-matching telescope and beam in order to maximize the cavity resonance. To minimize aberration, it's important that the beam goes through the center of the lenses in the telescope. (If the beam does go through the center, it will hit the same spot far downstream as it did before you inserted the lenses.) Once you are sure everything is aligned, play with the cavity and input beam alignment slightly so that you obtain resonances for a variety of low-order Hermite-Gaussian (TEM$_{mn}$) modes. If the cavity is very well aligned you may see symmetric combination of the TEM$_{00}$ modes such as the "donut mode": TEM$_{01}$ + TEM$_{10}$. Place a camera after the end mirror of the cavity so that it receives the leakage beam through this mirror. Capture examples of several resonant modes. Do fits to cross sections of the images you've obtained and explain any deviations from pure Hermite-Gaussians.

Recommendations: If the modes go by too fast for your camera to catch in video mode, try putting a small cardboard box over the cavity with holes cut out for the input and output beams. This may help with air currents and temperature changes. Be careful not to bump the cavity in the process.

Hook the PZT up to a function generator. Apply a slow, ~0.3 Hz, sawtooth voltage to the PZT. The amplitude of the sawtooth wave from the signal generator needs to be sufficient to scan the mirror over a distance of at least one half wavelength. While you scan the cavity length in this manner, use the camera that receives the leakage beam from the cavity to take a movie of the light transmitted through the end mirror. Show that the order of the TEM$_{mn}$ modes appears as expected from the Gouy phase of the modes. Finally, replace the camera with a photodiode and acquire the transmitted light power as a function of time as you scan much more quickly, at ~3 – 10 Hz, still making sure the scan amplitude is sufficient to cover at least one half wavelength, corresponding to one free spectral range. Explain the spacing of the higher-order mode transmission peaks in terms of the g-factor of your cavity. Should the relative amplitudes of higher-order modes depend on the cavity stability?

Ideas for Further Investigation

Align the cavity so that only spherically symmetric modes are obtained. Capture images of a few of these and compare them quantitatively to an alternate set of orthonormal functions known as Laguerre Gaussians. Like the Hermite-Gaussians, the Laguerre Gaussians form a basis for the modes of a cavity.

Use a beamsplitter to direct some known fraction of the beam reflected from the input mirror of your cavity to a photodiode. Use the photodiode to measure the power reflected from the cavity when the cavity is resonant and when it is non-resonant. (Drive the cavity slowly through resonance.) Use the information to characterize the level of modematching you've obtained. If you also measure the power transmitted by the cavity, what can you say about optical losses and the level of cavity coupling?

8 Polarization

8.1 Polarized Light

Electromagnetic waves traveling in vacuum or any linear medium[1] are transverse waves. The electric and magnetic fields are perpendicular to the wave's direction of travel and perpendicular to each other. The direction of travel of the wave is $\vec{E} \times \vec{B}$. The electric and magnetic fields are in phase with each other and the magnetic field and electric field vector amplitudes are related by the speed of light v in the medium

$$\vec{B}_0 = \frac{\hat{z} \times \vec{E}_0}{v}. \tag{8.1}$$

In vacuum or linear media then, one can always find the magnetic field from the electric field and vice versa so it's enough to specify one or the other. The usual choice is to specify the electric field. Therefore, the language of polarization always refers to the properties of the electric field vector. For example, a plane-polarized electromagnetic wave traveling in the $+z$ direction with polarization in the $\hat{y}$ direction would be written

$$\vec{E}(z,t) = \hat{y} E_0 \cos(kz - \omega t) \tag{8.2}$$

$$\vec{B}(z,t) = -\hat{x} B_0 \cos(kz - \omega t). \tag{8.3}$$

If the $\hat{y}$ direction corresponds to the vertical axis, we would call this a "vertically polarized" plane wave even though the magnetic field (which carries just as much energy as the electric field) is oriented horizontally.

In general, an electromagnetic wave consists of *two* polarization components that can be expressed with respect to a transverse set of basis vectors, such as $\hat{x}$ and $\hat{y}$. With respect to this basis, $E_x(z,t)$ is the x-component of polarization and $E_y(z,t)$ is the y-component. The general expression for the electric field of a plane wave traveling in the $\hat{z}$ direction incorporates both of these components

$$\vec{E}(z,t) = \hat{x} E_x(z,t) + \hat{y} E_y(z,t) \tag{8.4}$$

$$= \hat{x} u_x \cos(kz - \omega t - \phi_x) + \hat{y} u_y \cos\left(kz - \omega t - \phi_y\right). \tag{8.5}$$

where the fields of the polarization components have been expanded to correspond to plane waves. The amplitudes u_x and u_y are constant and the components may have different phase leads ϕ_x and ϕ_y.

[1] A linear medium is a substance described by making the following replacements in Maxwell's Equations in vacuo: $\epsilon_0 \to \epsilon$, $\mu_0 \to \mu$.

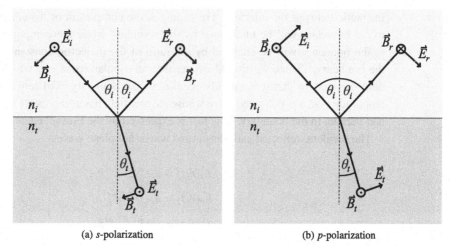

(a) *s*-polarization   (b) *p*-polarization

Figure 8.1 Reflection at an interface. Referring to Equations 8.6–8.8 at $\vec{r} = 0$, $t = 0$: $\vec{E_i}$ shows the field direction corresponding to positive values of u_i, $\vec{E_r}$ shows the direction corresponding to positive u_r, and $\vec{E_t}$ shows the field direction corresponding to positive u_t. The corresponding magnetic field directions $\vec{B}$ are also shown. (Although $\vec{r} = 0$ refers to the point of reflection/transmission at the interface, I've illustrated the field vectors some small distance away for clarity. To be fully correct though, the field vectors should be shown at the interface itself.)

When a linearly polarized beam is incident on a dielectric boundary, some of the beam will generally be reflected and some transmitted. In this situation, it's common to specify the polarization state relative to the plane of incidence. The plane of incidence is the plane made by the incident and reflected beam. If the electric field of the incident beam lies perpendicular to this plane, it is known as "*s*-polarized." If the electric field of the incident beam lies in the plane of incidence, the beam is known as "*p*-polarized."[2]

8.2 Fresnel Equations

One reason why it is important to know the polarization state of one's beams is that the reflective properties of materials are polarization dependent. For example, light that reflects from water at a shallow angle will be mostly horizontally polarized because the horizontal polarization is reflected better than the vertical one under those circumstances. Polarized sunglasses therefore cut down the glare by blocking horizontally polarized light.

The Fresnel equations describe the fraction of light reflected from a planar interface between transparent dielectric materials (air, water, glass, etc.). The situation is shown in Figure 8.1. The fact that the reflection is polarization dependent is a consequence of the boundary conditions on the parallel and perpendicular components of the electric and

---

[2] These designations come from German where "senkrecht" means perpendicular.

magnetic fields at the interface. For example, the component of the electric field parallel to the boundary will be unaffected by the boundary while the component perpendicular to the boundary will be changed by the ratio of the dielectric constants on each side of the boundary.[3] Thus, we should expect that an s-polarized wave, which has its electric field oriented so that it is entirely parallel to the boundary, will reflect with a different amplitude than a p-polarized wave whose electric field has components both perpendicular and parallel to the boundary. This fact is expressed by the Fresnel equations.

The incident, reflected and transmitted waves are plane waves

$$\vec{E}_i(\vec{r}, t) = u_i e^{-i\left(\vec{k}_i \cdot \vec{r} - \omega t\right)} \hat{n}_i \tag{8.6}$$

$$\vec{E}_r(\vec{r}, t) = u_r e^{-i\left(\vec{k}_r \cdot \vec{r} - \omega t\right)} \hat{n}_r \tag{8.7}$$

$$\vec{E}_t(\vec{r}, t) = u_t e^{-i\left(\vec{k}_t \cdot \vec{r} - \omega t\right)} \hat{n}_t, \tag{8.8}$$

where the $\vec{k}$s are the vector wave numbers ($|k| = \frac{2\pi}{\lambda}$, $\hat{k} \propto \vec{E} \times \vec{B}$) and the $\hat{n}$s are the polarization direction for each component (which remain oriented in the same plane). u_i, u_r and u_t are all real constants for a plane wave. For definiteness, we'll take $\vec{r} = 0$ at the point of reflection at the interface. A phase shift of 180° can occur upon reflection but not transmission, which means that assuming u_i is positive, u_t will also be positive and u_r can be either positive or negative. The amplitude and transmission coefficients r and t are defined as

$$r \equiv \frac{u_r}{u_i} \tag{8.9}$$

$$t \equiv \frac{u_t}{u_i}. \tag{8.10}$$

The Fresnel equations give r and t for both s and p polarization

$$r_s = \frac{n_i \cos\theta_i - n_t \cos\theta_t}{n_i \cos\theta_i + n_t \cos\theta_t}, \tag{8.11}$$

$$r_p = \frac{n_i \cos\theta_t - n_t \cos\theta_i}{n_t \cos\theta_i + n_i \cos\theta_t}, \tag{8.12}$$

$$t_s = \frac{2 n_i \cos\theta_i}{n_i \cos\theta_i + n_t \cos\theta_t}, \tag{8.13}$$

$$t_p = \frac{2 n_i \cos\theta_i}{n_t \cos\theta_i + n_i \cos\theta_t}. \tag{8.14}$$

Figure 8.1 shows the electric and magnetic field directions chosen to correspond to positive wave amplitudes.[4] n_i and n_t are the indices of refraction of the incident wave and transmitted wave mediums. θ_i and θ_t are the incident and transmitted angles respectively. To make

---

[3] See, for example, chapter 4 in Griffiths (2017).

[4] There is an alternative and fairly common sign convention for the positive direction of the reflected field for p-polarization that is opposite to that used here. An unfortunate effect of that convention is that the reflection coefficients for p- and s-polarization have opposite signs at normal incidence (where both polarizations behave in a *physically* identical manner). Under that sign convention, the right hand side of Eq. (8.12) gains an overall minus sign.

these equations useful in practice, where we usually only know the angle of incidence, we replace $\cos\theta_t$ using

$$\cos\theta_t = \sqrt{1 - \left(\frac{n_i}{n_t}\right)^2 \sin^2\theta_i} \tag{8.15}$$

obtained from $\cos^2\theta + \sin^2\theta = 1$ and Snell's law.

The corresponding ratios of the reflected or transmitted irradiance to the incident irradiance are indicated by capital letters, such as

$$R_p = \frac{\text{Reflected irradiance in } p\text{-polarization}}{\text{Incident irradiance in } p\text{-polarization}}. \tag{8.16}$$

Since optical irradiance is proportional to the square of the electric field amplitude, the corresponding irradiance ratios are just the squared amplitude ratios $R_p = r_p^2$, $T_p = t_p^2$, $R_s = r_s^2$, $T_s = t_s^2$. In the absence of optical losses, energy conservation requires

$$R_p + T_p = 1 \tag{8.17}$$
$$R_s + T_s = 1. \tag{8.18}$$

Example 8.1 Air↔Glass Interface   Perhaps the most important application of the Fresnel equations is to find the reflection from and transmission into a glass optic in air. While the surfaces of many optics have optical coatings on them, there are many cases where bare glass is used. The Fresnel equations apply to plane waves incident on planar surfaces. Locally, (for small enough cross-sectional areas) all beams are well approximated by plane waves and optical surfaces are flat. So, the Fresnel equations always apply locally. The Fresnel equations also give a good approximation to the average reflection coefficient over a whole beam whenever the beam's wavefronts are fairly flat and the angle of the optic's surface doesn't change significantly over the diameter of the beam.

The index of optical glasses varies according to the glass type. The index of common optical glasses at visible wavelengths ranges from about 1.45 for silica glass through 1.5–1.6 for crown glasses and up to around 1.7 for dense flint glasses. It's common to use $n = 1.5$ as a reference value for "the index of glass" when the glass type is unspecified and we adhere to that tradition in this example. The index of air is as close to one as makes no difference. The reflection and transmission coefficients are given by the Fresnel equations and shown in the figure below.

The solid lines correspond to rays passing from air into glass while the dashed lines correspond to rays passing from glass into air. The amplitude coefficients r_s and r_p are shown in the left panel. The irradiance (or power) coefficients R_s and R_p are shown in the right panel. Note that in the right panel, the scale of the y-axis has been expanded.

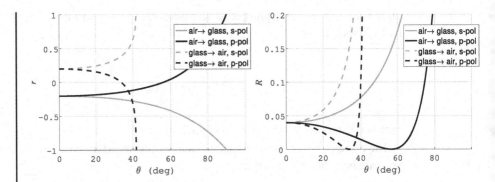

The following features are worth pointing out:

- The reflected *p*-polarization amplitude changes sign and passes through zero at the "Brewster" angle. At this angle, there is no reflected beam and the incident beam is transmitted in its entirety. For air-to-glass, the Brewster angle is 56.3°.
- For waves passing from low to high index (air→glass) at normal incidence, the reflected electric field at the interface points in the *opposite* direction to the incident electric field at the interface. (For a simple derivation of this I recommend Griffiths (2017), Section 9.3.2.) For this reason, the sign of the solid curves in the left panel above is negative at normal incidence. For waves passing from high to low index (glass→air) at normal incidence, the reflected electric field points in the *same* direction as the incident electric field. For this reason, the sign of the dashed curves in the left panel above is positive at normal incidence.

8.3 Jones Vectors

As in Chapter 1, Eq. (8.5) is usually expressed in "complex form." Generalizing to the complex plane simplifies the math needed to superpose waves efficiently and it simplifies the representation of polarizing optics. The complex form of Eq. (8.5) is

$$\vec{E}(z,t) = \hat{x}u_x e^{-i(kz-\omega t)} + \hat{y}u_y e^{-i(kz-\omega t)}. \tag{8.19}$$

The phase offsets of the polarizations, ϕ_x and ϕ_y are incorporated into the corresponding complex-valued amplitudes, $u_x = |u_x|e^{i\phi_x}$ and $u_y = |u_y|e^{i\phi_y}$.

The effect of a polarizing optic on the polarization state is described by the effect on the complex-valued polarization amplitudes, u_x and u_y. Polarizing optics can reduce the optical power in one or both polarization components, so to keep track of power changes, we choose some place in an optics chain (generally at the beginning) to serve as a reference. The overall amplitude of the beam at this reference point is $u_0 = \sqrt{|u_x|^2 + |u_y|^2}$. We normalize the individual polarization amplitudes by the reference amplitude u_0 giving

$$\vec{E}(z,t) = \left(\hat{x}\frac{u_x}{u_0} + \hat{y}\frac{u_y}{u_0}\right)u_0 e^{-i(kz-\omega t)} \tag{8.20}$$

$$= \left(\hat{x} J_x + \hat{y} J_y\right) u_0 \, e^{-i(kz-\omega t)} \tag{8.21}$$

$$= \vec{J} u_0 \, e^{-i(kz-\omega t)}. \tag{8.22}$$

J_x and J_y are the two components of the so-called Jones vector, $\vec{J}$, which specifies the polarization state of the beam. For example, a vertically ($\hat{y}$) polarized beam is described by the Jones vector $\vec{J} = \begin{pmatrix} 0 \\ 1 \end{pmatrix}$.

The Jones vector at the reference position in the optics chain has unit magnitude due to the normalization. $\vec{J}^* \cdot \vec{J}$ is proportional to the optical power so the Jones vector keeps track of power lost to polarizing optics. For example, if a polarizing optic has reduced the optical power by half as compared to the reference power, then $\vec{J}^* \cdot \vec{J} = \frac{1}{2}$.

The same format can be used when the electromagnetic wave is not a plane wave. For example, an arbitrary paraxial beam with its two polarization components is written

$$\vec{E}(x, y, z, t) = \vec{J} u(x, y, z) e^{-i(kz-\omega t)}. \tag{8.23}$$

The uniform amplitude u_0 in Eq. (8.22) has been replaced by the complex field amplitude $u(x, y, z)$. Equation (8.23) is a pleasingly compact way of describing an arbitrary paraxial beam.

A word of caution: The Jones vector formalism only applies to beams where all the photons are in the same polarization state. Partially polarized beams and beams containing polarization mixtures need more parameters to describe their state.[5]

Example 8.2 Electric Field of a Fully Polarized TEM$_{mn}$ Mode   This is a good time to write down the complete expression for the electric field of a TEM$_{mn}$ mode beam, including polarization. In Eq. (8.23), we substitute in for $u(x, y, z)$ from Eq. (4.13), obtaining

$$\vec{E}(x, y, z) = \vec{J} u_0 \frac{w_0}{w(z)} H_m\left(\frac{\sqrt{2}x}{w(z)}\right) H_n\left(\frac{\sqrt{2}y}{w(z)}\right) e^{-\frac{x^2+y^2}{w^2(z)}} e^{-ik\frac{x^2+y^2}{2R(z)}} e^{-i(kz-\omega t)+i\phi_0+i\theta_0}$$

$$\tag{8.24}$$

where the waist is assumed to be at the origin and all quantities are defined as before:

- z is the optic axis, x and y are the transverse coordinates,
- $k = \frac{2\pi}{\lambda}$ is the wavenumber, λ is the wavelength,
- $\omega = 2\pi f$ is the angular frequency, f is the frequency,
- $\vec{J} = \begin{pmatrix} J_x \\ J_y \end{pmatrix}$ is the Jones vector, and $\vec{J}^* \cdot \vec{J} = 1$,
- the amplitude $u_0 \in \mathbb{R}$ is constant,
- $w(z) = w_0 \left[1 + \left(\frac{z}{z_0}\right)^2\right]^{\frac{1}{2}}$ is the beam radius (amplitude radius),
- $R(z) = z\left[1 + \left(\frac{z_0}{z}\right)^2\right]$ is the wavefront curvature,

---

[5] Such beams may be described by a 4-element "Stokes vector." The effect of an optic is captured by multiplying the Stokes vector with a 4 × 4 "Mueller matrix." See, for example, Bass (2010), chapter 14.

- $\phi_0(z) = (1 + n + m) \tan^{-1}\left(\frac{z}{z_0}\right)$ is the Gouy phase,
- w_0 and R_0 are the beam radius and the wavefront curvature at the waist, which is at $z = 0$,
- H_n and H_m are the Hermite polynomials, the lowest order is $H_0 = 1$,
- $z_0 = \frac{\pi w_0^2}{\lambda}$ is the Rayleigh range,
- I've included an arbitrary constant phase lead θ_0.

8.3.1 Linear, Circular, and Elliptical Polarization

Figure 8.2 shows linear, circular, and elliptical polarization states labeled by the corresponding Jones vectors. We generally visualize polarization states by keeping track of the electric field vector in a plane as the wave travels perpendicularly through the plane. Since electromagnetic waves are transverse, the electric field vector always lies in this plane. If we plot the tip of the electric field vector in such a plane as a function of time, we can get a straight line, a circle or an ellipse. (See the lower panels of Figure 8.2.) If the tip of the field vector traces out clockwise motion (looking along the direction of travel of the wave $\vec{E}(\vec{r}) \times \vec{B}(\vec{r})$) then we say the beam is "right polarized." Counterclockwise motion is "left polarized." We could equally take a static "snapshot" of the field vector tip positions along an axis parallel to the direction of motion. (See the upper panels of Figure 8.2.) This static snapshot of tips of the field vector shows a simple sine wave for linear polarization and circular/elliptical spirals for circular/elliptical polarizations.[6]

8.4 Jones Matrixes

The effect of an optic on the polarization state is captured by multiplying the Jones vector $\vec{J}$ by a 2×2 Jones matrix, M to get a new Jones vector $\vec{J}'$.

$$\vec{J}' = M\vec{J}. \tag{8.25}$$

For example, the effect of a horizontal polarizer acting on a 45° linearly polarized beam would be described by the operation

$$\begin{pmatrix} J'_x \\ J'_y \end{pmatrix} = \begin{pmatrix} 1 & 0 \\ 0 & 0 \end{pmatrix} \begin{pmatrix} \frac{1}{\sqrt{2}} \\ \frac{1}{\sqrt{2}} \end{pmatrix} \tag{8.26}$$

$$= \begin{pmatrix} \frac{1}{\sqrt{2}} \\ 0 \end{pmatrix}. \tag{8.27}$$

The Jones matrix for a horizontal polarizer has the effect of removing the y-component of polarization but leaves the x-component untouched. Note that the square magnitude of

---

[6] The fact that the static snapshots for *left* circular/elliptical polarizations exhibit a *clockwise* spiral is surprising at first. But consider the motion of the field vector tip in the xy-plane as the spiral moves forward along the z-axis. A clockwise spiral leads to counter clockwise motion of the field vector tip in the xy-plane.

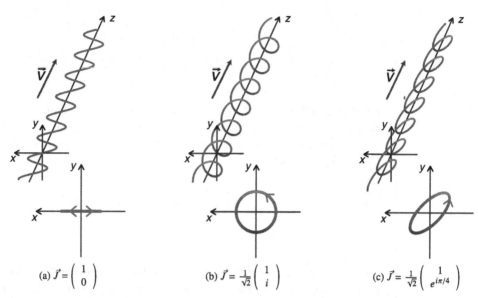

(a) $\vec{J} = \begin{pmatrix} 1 \\ 0 \end{pmatrix}$ (b) $\vec{J} = \frac{1}{\sqrt{2}} \begin{pmatrix} 1 \\ i \end{pmatrix}$ (c) $\vec{J} = \frac{1}{\sqrt{2}} \begin{pmatrix} 1 \\ e^{i\pi/4} \end{pmatrix}$

Figure 8.2 Three polarization states. These graphs show the progression of the tip of the electric field vector. The upper panels are "snapshots" at some arbitrary time. The lower panels show the motion in a plane at some particular z-position as time goes by. **(a)** Linear horizontal polarization. **(b)** Left-circular polarization.[6] **(c)** Left-elliptical.[6]

the Jones vector changes by the same fraction as the beam power. Before the polarizer $|\vec{J}|^2 = 1$, while after the polarizer $|\vec{J}'|^2 = \frac{1}{2}$. In other words, the polarizer reduces the beam power by 50%, as expected.

We can also find the Jones matrix of a polarizer that is *not* oriented so as to produce an x-polarized beam, but is rotated by some angle θ with respect to the x-axis. We obtain its Jones matrix using a coordinate rotation. We rotate our coordinate system by $-\theta$ so that in the new coordinate system our polarizer is oriented along the x-axis. Then we can apply the Jones matrix for a horizontal polarizer before rotating back to the original coordinate system. Using the rotation matrix $\Lambda(\theta)$ corresponding to coordinate rotations (passive rotations)

$$M' = \Lambda(\theta)^{\mathrm{T}} M \Lambda(\theta) \tag{8.28}$$

$$= \begin{pmatrix} \cos\theta & -\sin\theta \\ \sin\theta & \cos\theta \end{pmatrix} \begin{pmatrix} 1 & 0 \\ 0 & 0 \end{pmatrix} \begin{pmatrix} \cos\theta & \sin\theta \\ -\sin\theta & \cos\theta \end{pmatrix} \tag{8.29}$$

$$= \begin{pmatrix} \cos^2\theta & \cos\theta\sin\theta \\ \cos\theta\sin\theta & \sin^2\theta \end{pmatrix}. \tag{8.30}$$

For the purposes of Jones matrices, positive θ corresponds to optics rotated in a right-hand sense with respect to the beam direction (i.e. clockwise looking downstream). As expected, if one rotates a horizontal polarizer through $\theta = 90°$, the result is a vertical polarizer.

$$M' = \Lambda(90°)^{\mathrm{T}} M \Lambda(90°) \tag{8.31}$$

$$= \begin{pmatrix} 0 & -1 \\ 1 & 0 \end{pmatrix} \begin{pmatrix} 1 & 0 \\ 0 & 0 \end{pmatrix} \begin{pmatrix} 0 & 1 \\ -1 & 0 \end{pmatrix} \qquad (8.32)$$

$$= \begin{pmatrix} 0 & 0 \\ 0 & 1 \end{pmatrix}. \qquad (8.33)$$

Example 8.3 Malus' Law   We can use Eq. (8.30) to find the effect of a linear polarizer at an angle θ with respect to the polarization direction of a linearly polarized beam. We take the initial Jones vector for a horizontally ($\hat{x}$) polarized beam and multiply it by the matrix 8.30.

$$\vec{J}' = \begin{pmatrix} \cos^2 \theta & \cos \theta \sin \theta \\ \cos \theta \sin \theta & \sin^2 \theta \end{pmatrix} \begin{pmatrix} 1 \\ 0 \end{pmatrix}. \qquad (8.34)$$

$$= \begin{pmatrix} \cos^2 \theta \\ \cos \theta \sin \theta \end{pmatrix}. \qquad (8.35)$$

So, $|\vec{J}'|^2 = \cos^4 \theta + \cos^2 \theta \sin^2 \theta = \cos^2 \theta$. Thus the irradiance I' of a linearly polarized beam after passing through a polarizer at an angle θ with respect to the original polarization direction of the beam is

$$I' = I \cos^2 \theta, \qquad (8.36)$$

where I is the original irradiance. This relationship is known as "Malus' law,"

The Jones matrixes for three common polarizing optics are shown in Table 8.1. The effect of multiple polarizing optics in succession is described by multiplying the Jones matrixes for the individual optics. Like ABCD matrix products, the Jones matrix product is performed in reverse order; the first optic encountered is the last (rightmost) matrix in the product.

8.4.1 Optical Isolators

Sometimes it is necessary to block "rogue beams" traveling backward up an optical train due to retroreflection from optical surfaces. Such reflections can cause "parasitic interference" between any two optics and are likely to pollute output beams. When such rogue beams reenter a laser they can also increase the amplitude and/or frequency noise on the beam. The simplest way to reduce the number of rogue beams is to "misalign" transmissive optics slightly so that retro reflections aren't coincident with the original beam and end up getting dumped on optic holders, housings, and so forth.

Sometimes intentional misalignment is not possible such as when aligning an external cavity. It may also be undesirable for other reasons such as when sensitivity to scattered light is high or when working with high-power lasers. In such cases, counter-propagating beams must be blocked using an optical isolator, which acts like a "diode for light." It allows beams to pass in one direction but blocks beams traveling in the opposite direction. The most commonly used optical isolator is a Faraday isolator. A Faraday isolator uses the so-called Faraday effect to rotate a beam by 45° between two polarizers rotated by 45° with respect to one another. The sense of the rotation depends on the direction of travel of

Table 8.1 Jones matrixes for polarizing optics. To obtain the Jones matrixes corresponding to rotated optics, use these in conjunction with Eq. (8.28). Any additional phase lag $\Delta\phi$ imposed on the beam by the increased optical pathlength of the optic as compared to free space can be incorporated by multiplying these matrixes by $e^{i\Delta\phi}$. Such phase lag applies to both polarization components equally.

Description	M	Comments
Linear polarizer, $\hat{y}$-polarization	$\begin{pmatrix} 0 & 0 \\ 0 & 1 \end{pmatrix}$	For arb. angle see Equation (8.30).
Waveplate, fast-axis $\parallel \hat{y}$	$\begin{pmatrix} 1 & 0 \\ 0 & e^{i\psi} \end{pmatrix}$	ψ is the phase lead of the fast axis compared to the slow axis.
Rotator, Right-handed	$\begin{pmatrix} \cos\beta & -\sin\beta \\ \sin\beta & \cos\beta \end{pmatrix}$	Rotation angle β is clockwise looking downstream when $\vec{B}$ points downstream.

the beam. Thus, a Faraday isolator allows beams traveling in one direction to pass while blocking beams traveling in the opposite direction.

Sometimes, it's easier to use a simpler isolator consisting of a polarizer followed by a quarter-wave plate. Consider a linear polarizer oriented at 45° with respect to the vertical followed by a quarter-wave plate whose fast axis is vertical. The idea is that the beam leaving this combination is circularly polarized due to the $\frac{\pi}{2}$ phase lead induced onto the y-component of the field. If a portion of this circularly polarized beam is retroreflected up the beam, the quarter-wave plate will impart another $\frac{\pi}{2}$ phase lead onto the y-component making the total phase lead of that component π radians. That's just the same as flipping the sign of the y-component of the field. This causes the original 45° polarization to change to -45°. Such a polarization can't pass through the polarizer and the retroreflected light is blocked. The Jones matrix seen by such a retroreflection is

$$M = \overbrace{\frac{1}{\sqrt{2}}\begin{pmatrix} 1 & 1 \\ 1 & 1 \end{pmatrix}}^{45° \text{ polarizer}} \overbrace{\begin{pmatrix} 1 & 0 \\ 0 & i \end{pmatrix}}^{\frac{\lambda}{4} \text{ plate}} \overbrace{\begin{pmatrix} 1 & 0 \\ 0 & i \end{pmatrix}}^{\frac{\lambda}{4} \text{ plate}} \overbrace{\frac{1}{\sqrt{2}}\begin{pmatrix} 1 & 1 \\ 1 & 1 \end{pmatrix}}^{45° \text{ polarizer}} \qquad (8.37)$$

$$\overbrace{= \frac{1}{2} \begin{pmatrix} 1 & 1 \\ 1 & 1 \end{pmatrix} \begin{pmatrix} 1 & 0 \\ 0 & -1 \end{pmatrix}}^{= \frac{4}{2} \text{ plate}} \begin{pmatrix} 1 & 1 \\ 1 & 1 \end{pmatrix} \tag{8.38}$$

$$= \begin{pmatrix} 0 & 0 \\ 0 & 0 \end{pmatrix}. \tag{8.39}$$

In other words, the resulting Jones matrix is the zero matrix that will take any incident Jones vector to zero. This whole example could equally have been done with the system rotated so that the polarizers are oriented along a coordinate axis and the waveplate is at 45°. The result is of course the same.

For this type of isolator, sometimes called a "free-space circulator," it's important that optics downstream of the isolator don't change the polarization state. In other words, there can't be an extra Jones matrix in between the two quarter-wave plate matrixes. If such a matrix were present, the product of the four matrixes would no longer be zero and light could pass up through the isolator. Note also that the polarization state in the optics chain downstream of the isolator is circular, which may be inappropriate for some experiments. One reason for the popularity of Faraday isolators, is that they work regardless of the polarization state of the retroreflected beam. However, they are also generally more expensive and always incorporate a strong magnet that could be an issue in some cases.

Exercises

8.1 By appealing to the equivalent formulae for s and p polarized light, show that an arbitrary linear polarization satisfies $R + T = 1$ where

$$R = \frac{\text{Total reflected irradiance}}{\text{Total incident power}}$$

and

$$T = \frac{\text{Total transmitted irradiance}}{\text{Total incident power}}.$$

8.2 Show that Eq. (8.12) admits an angle of incidence, called the Brewster angle $\theta_i = \theta_B$, at which the reflectivity is zero and all incident light is transmitted. Derive an expression for θ_B and find its value in degrees to three significant digits for an air-to-water interface at mid-visible wavelengths: $n_{air} = 1.00$, $n_{water} = 1.33$.

8.3 From Eq. (8.11) show that no Brewster angle exists for s–polarized light.

8.4 Use the Fresnel equations and Snell's law to describe a fish's view through the glass of the environment outside its fishtank.

8.5 Uncoated transmissive optics are usually less expensive than the corresponding anti-reflection coated versions. Typical commercial antireflective coatings may reduce the reflection from each surface of a lens to about 0.2%. In an optics chain involving ten transmissive optics, how much of the input power can you expect to lose when using uncoated glass optics versus antireflection-coated optics?

8.6 Show that a linearly polarized plane wave, polarized at an angle α with respect to the x–axis has Jones vector

$$\vec{J} = \begin{pmatrix} \cos \alpha \\ \sin \alpha \end{pmatrix}.$$

Hint: $\vec{J}^* \cdot \vec{J} = 1$.

8.7 A half-wave plate causes the fast axis polarization component to lead the slow axis component by π radians. The Jones matrix for this optic, oriented so that the fast axis is aligned with the y–axis, is

$$M = \begin{pmatrix} 1 & 0 \\ 0 & -1 \end{pmatrix}.$$

Explain why this is the correct Jones matrix.

8.8 Find the Jones matrix of two linear polarizers in sequence. The polarizers are rotated by 45° with respect to one another. Calculate the transmissivity $T(\theta)$ of this system for incident light that is linearly polarized with polarization angle θ with respect to the first polarizer. (The transmissivity is the fraction of the power passed through the system.) Plot $T(\theta)$ for $0 \le \theta < 2\pi$.

8.9 Show that an optic described by a Jones matrix with purely real components and acting on linearly polarized light, can only produce linearly polarized output.

8.10 Find the Jones matrix of quarter-wave plate with the fast axis rotated 45° counter-clockwise from the vertical. If this plate acts on a vertically polarized beam, show that circular polarization results.

8.11 Consider two beams that are combined at a beamsplitter. Argue that the Jones vectors of the two beams should be linearly combined to produce the overall Jones vector of the combined beam.

8.12 Find the Jones vector of the output beam at the antisymmetric port of a Michelson interferometer with s-polarized input light. The x-arm has no polarization effect but the y-arm is slightly birefringent. The y-arm mirror acts as a waveplate with a fast axis at 45° to the input polarization and phase lead $\psi = \epsilon \ll 1$. In terms of ϵ, find the fringe contrast (use Eq. (5.13)). *Hint*: Like the electric fields themselves, the Jones vectors satisfy linear superposition.

8.5 Experiment: Investigation of Polarized Light

Objectives

1 Observe naturally occurring polarization states including partially polarized reflected light and scattered sunlight.
2 Verify the Fresnel equations.
3 Verify Malus' law.
4 Relate your observations of the light passed by three polarizers in sequence to wave-function collapse in quantum mechanics.
5 Compare the behavior of a quarter-wave plate to theory.
6 Build and characterize an optical isolator.

Equipment Needed

- Two sheets of polarizing film (or two polarizing plates) through which to view the sky and other everyday objects. It's convenient if these are 5 cm per side or larger.
- Linearly polarized laser. (A laser with random polarization passed through a polarizer will work provided the power in the chosen polarization does not fluctuate significantly.)
- A pair of fixed linear polarizers with orthogonal polarization axes.
- One rotatable polarizer in a graduated rotation mount.
- Amplified photodiode.
- Uncoated glass wedge or an uncoated plano-convex lens. These should be set in a mount that can rotate about the vertical axis and which has a scale to read the rotation angle to at least 2° accuracy.

Polarization in the Environment

Using two sheets of polarizing film, look at an unpolarized light source like a light bulb. First use one sheet and rotate it to different angles. Make sure the source is not polarized. Now add the second sheet and rotate it with respect to the first sheet. Describe what you see.

Next, use a single sheet of polarizing film to investigate the polarization state of the following sources and comment on the qualitative features:

- The light from an LCD monitor. What is the purpose of the observed polarization?
- A reflected image in a window, especially with high angle of incidence.
- The blue sky (if conditions permit). Estimate the direction of the dominant polarization at some position on the sky. Draw a diagram/map of the sky illustrating how the dominant polarization direction varies with sky position. Use a line to indicate the axis of the dominant polarization. The length of the line should indicate the relative strength of the dominant polarization. Make sure you indicate the sun's location on your diagram.
- The cloudy sky (if conditions permit). Look for any polarization features.

Note that most naturally polarized light is only partially polarized. In other words, a larger fraction of photons may be in one polarization state than the other but the light is far from being uniformly polarized. When you look at such partially polarized light through a

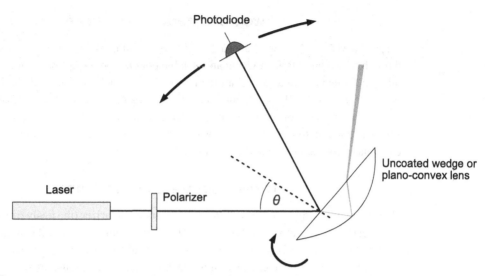

Figure 8.3 Possible setup for testing the Fresnel equations.

polarizing sheet, the amount of light transmitted through the sheet will depend on the angle of the sheet but will never be fully extinguished.

Fresnel Equations

Working at an optical table, test the Fresnel equations for both s-polarization and p-polarization. Do this by measuring the fraction of incident power reflected from a glass surface. Fit your data for each polarization to the expected functional form using the index of refraction, n, as a fit parameter. Find the uncertainty in n. Generate the reduced chi-square and decide whether your model adequately explains your data. If it does not, make sure you've accounted for any systematics.

> *Recommendations*: You will use a photodiode to measure both the incident and the reflected light power. Make sure the reflected beam falls entirely onto the active area of the photodiode and no light spills off the edge. If the output of the laser you are using is a single linear polarization state, you will need to rotate the laser to achieve the desired polarization direction (s or p). If the output of the laser you are using is not a single linear polarization state, you will need to polarize the light in the appropriate direction using a standalone polarizer. It will be necessary to isolate the front surface reflection so as not to confuse it with the back surface reflection. This is most easily done by using a wedge that will cause the back surface reflection to leave the sample at a different angle. An alternative is to use a short focal length plano-convex lens and make sure the beam hits the lens off-center. (See Figure 8.3.)

Two Polarizers in Sequence

Prepare a vertically polarized beam, either by rotating your laser or passing the beam through a vertical polarizer. Using an rotatable polarizer as an analyzer, measure the transmission through the analyzer as a function of analyzer polarization angle θ. (The analyzer polarizer should already be pre-mounted in a rotating optical holder with a scale to indicate θ.) Take $\theta = 0$ to be parallel to the input polarization. Fit or compare the transmitted power versus angle data to the functional form expected from Malus' law. Based on the reduced chi-square, comment on whether or not Malus' law is correct.

Three Polarizers in Sequence

Now place an "output" polarizer *after* the adjustable polarizer (the analyzer). The polarizing direction of the output polarizer should be perpendicular to that of the input polarization so that if the analyzer is removed, no light is transmitted. Now follow the same procedure as in the two-polarizer case, recording the fractional power transmitted as a function of analyzer angle. Fit the data to your best guess for the functional form. It may surprise you that any light is transmitted at all. If no input light is in the polarization state passed by the output polarizer, how can the analyzer change the light so that some of it is in the state of the output polarizer? This is in fact a quantum mechanical effect. The analyzer collapses the vertical-polarization photon wavefunctions onto the polarization direction of the analyzer. These new wavefunctions now have some overlap with the horizontal polarization state passed by the output polarizer, thereby allowing some light to pass through. Without the intermediate collapse of the wavefunction due to the analyzer, no light can pass through the orthogonal input and output polarizers.

Quarter-Wave Plate

As discussed in Chapter 2, a quarter-wave plate has slightly different indices of refraction for two orthogonal polarizations. Thus, light of one polarization will travel faster in the wave plate than the other and accumulate less phase. The thickness of the quarter-wave plate material is chosen so that the polarization corresponding to the slow axis of the quarter-wave plate will lag the polarization corresponding to the fast axis of the plate by a phase angle of 90°. Passing linearly polarized light through a quarter-wave plate results in elliptically polarized light unless the input polarization is aligned with one of the axes of the quarter-wave plate. Place the fast axis of the quarter-wave plate at 45° relative to the input polarization. Experimentally, show that the output light is not linearly polarized. Show also that the fattest polarization ellipse results when the quarter-wave plate is at 45°. Finally, build and quantify the efficacy of the isolator whose action is described by Eq. (8.37).

Ideas for Further Investigation

You found that the Rayleigh scattering from the blue sky is polarization dependent. Find the ratio of the irradiance of light polarized along an axis passing through the sun to the total irradiance from that point in the sky. Do the same for light polarized along the perpendicular axis. How much does it vary across the sky? Make a *quantitative* sky-map of the polarization ratios.

Some materials are naturally birefringent. One such material is one of the most common minerals around, namely, calcite. When calcite occurs in large transparent crystals it is known as Iceland spar. Calcite treats the two polarizations of light differently as you can easily see by examining text through a piece of Iceland spar. Investigate and explain this behavior.

9 Optical Noise

9.1 Characterizing Noise

When considering new experiments, we do "back-of-the-envelope" calculations to see if our ideas are plausible. Very often, new experiments involve the detection of small signals. If we can estimate the expected signal and noise levels in the potential data stream from a new experiment, then we can generate the expected signal-to-noise ratio (SNR). The SNR is just the ratio of the power in the signal to the power in the noise at any given time. Roughly speaking, if the SNR is greater than one, then we have a chance of discerning the signal against the noise and the experiment may be worth doing, otherwise, not.

We go through a sample SNR calculation in Example 9.1. While signals take on forms specific to each experiment, technical noise sources tend to be ubiquitous. Fundamental noise sources, driven by the unchanging physics of light and materials are also present in all experiments. Because they are omnipresent, noise sources are worth studying in more detail and this chapter serves as a short introduction.

Often but not always, noise is a random fluctuation in the data stream such that if one plots a histogram of the data stream, the result is a Gaussian. Such noise is usually called Gaussian noise. An example of this is shown in Figure 9.1. The data stream is a random voltage $V(t)$ taken from a Gaussian parent distribution and sampled at uniformly spaced time intervals $\Delta t = 10^{-4}$ s, a sampling rate $f_s = \frac{1}{\Delta t} = 10^4$ Hz. The result is similar to the background noise seen on a digital or analog oscilloscope without any inputs connected and the gain turned all the way up. Different ways of viewing the data stream, $V(t)$, are shown in Figure 9.1. The time domain is a good way of viewing signals of wide bandwidth and short duration. The frequency domain, here represented by the power spectral density (PSD), is especially good for viewing signals of narrow bandwidth and long duration.[1]

9.1.1 Power Spectral Density

The power spectral density of a data stream tells us how much power the data stream contains per unit frequency at any particular frequency. In other words, it tells us how the power in the data stream is distributed among the frequencies. Technically, it is the Fourier transform of the data stream's autocorrelation and normally also equal to the squared

---

[1] Signals with medium bandwidth and medium duration may need specialized tools such as a time-frequency plot (also known as a spectrogram), wavelet analysis, matched filters, and so forth, all of which are beyond the scope of this book.

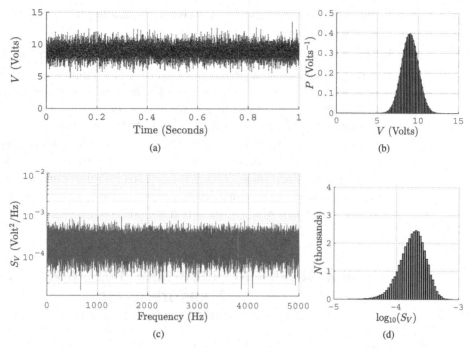

Figure 9.1 A simulated Gaussian random noise signal, sample rate $f_s = 10$ kHz viewed in four different ways. **(a)** The raw data stream is a time-domain voltage $V(t)$, **(b)** Probability density P of the voltage $V(t)$ with a superimposed fit to a Gaussian, **(c)** The power spectral density $S_V(f)$ of the voltage $V(t)$. **(d)** Histogram of the logarithm, base 10, of the power spectrum. *Notes*: **1.** The total length of the data stream is 36 seconds but only 1 second is shown in panel (a) so the internal structure can be seen. In that panel, black dots correspond to the actual data point locations. **2.** When viewing power spectra on a log scale, as in panel (c), it's a good idea to remember that the histogram of the power spectrum is *not* symmetric about the mean, which is clear from the histogram in panel (d). On a log scale, the mean value is lower than the most likely value, so the best estimate of the power spectral density is somewhat above the middle of the "scruff."

magnitude of the data stream's Fourier transform. Like the Fourier transform, the PSD involves an integral over an infinite time duration. Since real signals have finite duration, we are always dealing with "estimates of the true PSD." How to make those estimates is a subject of signal processing theory and several methods exist. For our current purposes however, we only need a conceptual understanding of the power spectrum and we'll simply rely on instruments or software tools to provide good power spectrum estimates for our data. There are four main things to remember about PSDs:

1. If the data stream $V(t)$ has units of Volts, the PSD $S_V(f)$ has units of $\frac{\text{Volts}^2}{\text{Hz}}$. $S_V(f)$ is positive and real and represents the power in a 1 Hz band centered at frequency f.

2. The maximum frequency in the PSD is half the sampling frequency and is known as the Nyquist frequency.

3. The duration T of the PSD sets the frequency resolution, Δf, according to: $\Delta f = \frac{1}{T}$.

4. A data stream consisting of only Gaussian noise of duration T can be split into N segments of duration T/N. The mean power spectral density of the N segments will have its RMS fluctuation reduced by a factor of $\sqrt{N}$ compared to a single power spectrum of duration T. Thus, averaging can reduce the PSD noise. This comes at the expense of a reduction in the frequency resolution by N. So, performing 100 averages will reduce the "fuzziness" of the spectrum by a factor of 10 at the expense of a 100-fold reduction in the frequency resolution. If spectral resolution isn't important, more averaging may be appropriate and vice versa. (Usually one wants some averaging. When I make power spectra, I often start with $N = 10$ and adjust from there.)

9.2 Quantum Fluctuations in Optical Power Measurements

In this section we introduce one of the quantum effects of light known as shot noise. Typical experiments using optics operate with very large[2] rate of photons passing through the apparatus and in this limit the photon nature of light tends to show up as shot noise. Shot noise is quite small and can often be ignored. Now we choose to investigate it for its own sake.

At the most fundamental level, the electric and magnetic fields constituting light are described by quantum field theory. The quantum description and statistical properties of photons are covered by the field of quantum optics. Quantum optics, which could be said to have originated in the early part of last century, grew enormously in the later part of that century and more again in the current one. One reason for this growth is simply the advent and development of the laser and associated tools that allowed physicists to construct experimental apparatus that was not only sensitive to quantum properties of light but was also able to manipulate them. A second reason for growth was that the quantum field theory framework needed to properly describe the quantum nature of light that became available. A proper introduction to quantum optics warrants at least a semester-long course, usually at the graduate level. We will quote the most basic results from that theory, and only as they relate to shot noise.

According to the quantum optical description, the quantum fields leaving a single-mode laser are in so-called coherent states $|\alpha>$ and these states are in turn superpositions of Fock states $|n>$, which are states of well-defined photon number.

$$|\alpha> = \sum_n a_n |n> . \tag{9.1}$$

---

[2] One picowatt of light at $\lambda = 633$ nm corresponds to about three million photons per second.

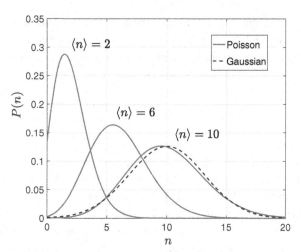

The continuous Poisson distribution for the mean values $\langle n \rangle$ indicated. A Poisson distribution and a Gaussian distribution are shown for $\langle n \rangle = 10$. The two distributions coincide ever more closely as $\langle n \rangle$ increases.

The properties of the coherent state are determined by the coefficients a_n, which are given by

$$a_n = \exp\left(-\frac{|\alpha|^2}{2}\right)\frac{\alpha^n}{\sqrt{n!}}, \tag{9.2}$$

α is a complex number with the property that it's squared magnitude is the expectation value of the number of photons in the state $|\alpha>$

$$\langle n \rangle = |\alpha|^2. \tag{9.3}$$

Any measurement of the number of photons in the field leaving the laser will have as a result an integer value n with probability $p(n) = |a_n|^2$. Since energy is proportional to photon number, a detector of optical energy such as a photodiode serves to collapse the coherent state onto one of the Fock states. Subsequent measurements of optical fields, newly arrived at the photodetector, will be uncorrelated and the number of photons measured in each unit of time will fluctuate randomly with the distribution specified by $|a_n|^2$. The distribution is a Poisson distribution.

$$p(n) = |\alpha_n|^2 = \langle n \rangle^n \frac{e^{-\langle n \rangle}}{n!}. \tag{9.4}$$

The corresponding fluctuation in the number of photons received as a function of time is known as shot noise.

The Poisson distribution for several low values of $\langle n \rangle$ is shown in Figure 9.2 and becomes indistinguishable from a Gaussian for large $\langle n \rangle$. So, shot noise is a form of Gaussian noise as depicted in Figure 9.1. The standard deviation of the distribution for a Poissonian process like this is

$$\sigma_n = \sqrt{\langle n \rangle}. \tag{9.5}$$

For coherent states, the power spectral density of fluctuations in the measured photon rate, $\dot{n}$, is white due to the fact that repeated measurements of the incoming light field are uncorrelated. The standard deviation of the Poisson distribution mentioned earlier sets the level of the power spectral density at

$$\boxed{S_{\dot{n}}(f) = 2\langle \dot{n}\rangle} \tag{9.6}$$

where $\langle \dot{n}\rangle$ is the average measured rate of photon arrival. The factor of two corresponds to the fact that we prefer to use single-sided power spectra having only positive frequencies. The single-sided amplitude is twice the double-sided version due to the negative frequency contributions being "folded over" and added to the, equal power, positive frequency ones. To reiterate, this equation is a consequence of the Poisson distribution inherent to the coherent states of light and the uncorrelated nature of repeated measurements. To convert this to the PSD of the *relative* photon flux, we just divide by $\langle \dot{n}\rangle^2$. The quantity is squared because $S_{\dot{n}}$ is the *power* spectral density of the quantity $\dot{n}$. (Power is always proportional to the amplitude squared.) Similarly, to convert Eq. (9.6) to PSD of optical power rather than photon rate, we need to multiply by the square of the energy per photon, $\left(\frac{hc}{\lambda}\right)^2$.

$$S_P(f) = 2\left(\frac{hc}{\lambda}\right)^2 \langle \dot{n}\rangle \tag{9.7}$$

$$= \frac{2hc}{\lambda}P. \tag{9.8}$$

Strictly speaking, we should use $\langle P\rangle$ to refer to the average optical power. However, we'll drop the $\langle\rangle$ notation since we're no longer talking about photons and quantum states. Finally, if we want to convert to relative power fluctuations, we need to divide by P^2 and since the relative power fluctuations and relative irradiance fluctuations are identical, we refer to the PSD of relative irradiance fluctuations as the "relative intensity noise," using the old-fashioned word for irradiance. The relative intensity noise (RIN) is a very common way of viewing noise. For shot noise it is

$$S_{\text{RIN}}(f) = \frac{2hc}{\lambda}\frac{1}{P}. \tag{9.9}$$

The units of relative intensity noise are Hz^{-1}.

Now that we've found the shot noise with coherent states, we should also mention that it's possible, with clever techniques in nonlinear optics, to prepare light in a so-called squeezed state where the distribution of photon numbers is *narrower* than Poissonian, thus reducing the shot noise.[3] However, the uncertainty principle extracts a toll by increasing the distribution of phase, thus increasing the phase noise. Similarly, the phase distribution

---

[3] For an excellent and in-depth introduction to quantum optics, I recommend *A Guide to Experiments in Quantum Optics*, Bachor and Ralph (2019). It covers the theory of squeezing and many other aspects of quantum optics from an experimentalist's point of view. It relates theory directly to actual laboratory observations and helps to build intuition into a subject that can otherwise seem rather impenetrable.

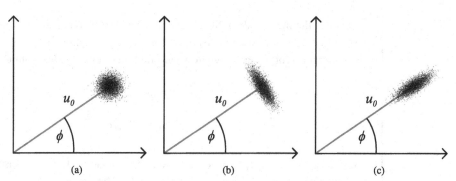

Figure 9.3 **(a)** Coherent state (unsqueezed), **(b)** Amplitude squeezing, **(c)** Phase squeezing.

can be squeezed at the expense of the amplitude distribution, causing increased shot noise. One way of illustrating this is to plot the phasor of the light field. It is a rapidly rotating arrow in the complex plane due to the $e^{i\omega t}$ component of the field. If we move to a set of co-rotating axes, the overall phase offset of the light is the angle from the x–axis and the amplitude of the light field is given by the length of the arrow. The uncertainty in the amplitude and phase can be illustrated by a cloud of dots whose density represents the probability of measuring the phasor's tip there. (See Figure 9.3.) The transverse width of the ellipse shows the phase uncertainty and the longitudinal width of the ellipse shows the amplitude uncertainty and corresponds to shot noise. The area of the ellipses is constant (assuming minimum uncertainty states).

Squeezed light sees application in several areas. In gravitational wave detectors, squeezed light is used to reduce the readout noise in the gravitational wave channel. Squeezed states are also used for experiments in quantum mechanics: quantum teleportation, quantum computing, and so forth. A significant difficulty to the use of squeezed light is the fact that even small optical losses tend to reduce the amount of squeezing and return the light to a normal coherent state.

Example 9.1 Shot Noise in a Michelson Interferometer   We can do a back-of-the-envelope calculation to find the limits shot noise places on the sensitivity of a simple Michelson interferometer. To make it interesting, we'll use the calculation to see whether a simple Michelson interferometer with no frills is capable of detecting gravitational waves. We shall do the calculation for a Michelson interferometer whose arms are 4 km long, like the longest current gravitational wave detectors, and we'll assume we have a 100 W shot noise limited laser similar to the ones used in current gravitational wave detectors. (See Section 5.2.2.) Such a Michelson, operating around the center of the fringe, would have 50 W exiting the antisymmetric and symmetric ports at all times. The PSD of shot noise in terms of photon arrival rate at the antisymmetric port photodetector is given by Eq. (9.6). The corresponding noise in the photodetector's photocurrent S_i is

$$S_i = 2 \langle \dot{n} \rangle e^2 \tag{9.10}$$

$$= 2ei \tag{9.11}$$

where e is the magnitude of the charge on an electron and i is the average photocurrent. (As in Eq. (9.8), we could be using $\langle i \rangle$ for the average photocurrent but we're dropping the $\langle \rangle$ notation for averages when we are no longer talking about photon numbers.)

$$i = \langle \dot{n} \rangle e \tag{9.12}$$

$$= \left(\frac{\lambda}{hc} \right) P_{\text{out}} e, \tag{9.13}$$

where P_{out} is the average power at the antisymmetric port, λ is the laser wavelength, h is Planck's constant and c is the speed of light. The quantity $\frac{\lambda e}{hc}$ is the charge liberated per unit optical energy. For simplicity, I'm assuming that the photodetector emits one electron for each photon that impinges on it – a quantum efficiency of one. (This is a reasonable approximation, since the best infrared photodiodes have quantum efficiencies above 0.9.) Assuming the Michelson is operating mid-fringe, $P_{\text{out}} = P_{\text{in}}/2$.

The shot noise is frequency independent in the signal band of a simple Michelson interferometer and so the total noise power contributed to an integrated signal will depend only on the effective bandwidth of the signal B, assuming we can reject all noise not in the band of the signal. If the time spent by the signal in the signal band of the detector is τ and the signal is perfectly well modeled (i.e., we know its shape) then $B = 1/\tau$. This represents the best-case scenario. In that case, since the PSD times the bandwidth is the total power,

$$\text{Noise power} = S_i/\tau \tag{9.14}$$

$$= \frac{e^2 \lambda P_{\text{in}}}{hc\tau}. \tag{9.15}$$

(The actual photocurrent power in Joules is proportional to this quantity by the inverse transimpedance of the photodiode amplifier. Since we'll end up taking a ratio with the signal power, we don't need to worry about that because the signal has the same factor of proportionality.) We'll assume that a strong gravitational wave maintains an RMS strain amplitude of $h_{\text{rms}} \approx 1 \times 10^{-23}$ for the duration of it's detected presence. We multiply by the arm length $L = 4$ km to get the corresponding arm length change. We also need to multiply by two to get the differential arm length change since the strain is present in both arms with opposite sign. Then we apply Eq. (5.10) to get the RMS optical power fluctuation at the detector and convert to the corresponding RMS photocurrent fluctuations i_{rms} using the charge liberated per unit optical energy, $\frac{\lambda e}{hc}$.

$$\text{Signal power} = i_{\text{rms}}^2 \tag{9.16}$$

$$= \left(\frac{4\pi e L P_{\text{in}}}{hc} h_{\text{rms}} \right)^2, \tag{9.17}$$

where the proportionality factor is the same as for the noise power. Don't confuse h (Planck's constant) and h_{rms} (the gravitational wave strain amplitude). Finally, we get the signal-to-noise ratio in terms of RMS strain amplitude by taking the square root of the ratio of the signal and noise powers

$$\sqrt{\frac{\text{Signal power}}{\text{Noise power}}} \approx \left(\frac{4\pi}{\sqrt{hc}}\right) \sqrt{\frac{P_{\text{in}}\tau}{\lambda}} L h_{\text{rms}} \tag{9.18}$$

$$= 4\pi \left(\frac{L}{\lambda} \sqrt{\langle \dot{n} \rangle}\right) \left(h_{\text{rms}} \sqrt{\tau}\right) \tag{9.19}$$

$$\approx 0.01. \tag{9.20}$$

In the spirit of our quantum-optical discussion earlier, the second line expresses the result in terms of the average number of photons entering the interferometer per unit time $\langle \dot{n} \rangle$. There the factor of merit of the gravitational wave antenna is clear: $\frac{L}{\lambda} \sqrt{\langle \dot{n} \rangle}$. Similarly, the factor of merit for the signal is: $h_{\text{rms}} \sqrt{\tau}$.

Putting in the numbers, we see that the signal-to-noise ratio is only about 1% and there's no chance of detecting even fairly strong gravitational waves with a simple Michelson interferometer with the parameters considered. That is why all current interferometric gravitational wave detectors have some form of optical cavities in the arms. Indeed, the interferometer is itself made into an optical cavity within which the arm-cavities reside. This is done using a "power recycling" mirror at the symmetric port and a "signal recycling" mirror at the antisymmetric port forming nested cavities enveloping the arm cavities. The resulting configuration is called a "dual recycled, Fabry-Perot Michelson" interferometer. The optical cavities serve to boost the signal by increasing the time photons, and particularly photons carrying a signal, spend in the instrument by a good three orders of magnitude. The accumulated phase from any passing gravitational wave is boosted by the same factor. So, the signal-to-noise ratio for a strong gravitational wave goes from 0.01 to 10 or more.

9.3 Technical Sources of Optical Noise

In addition to the fundamental optical noise due to the probabilistic nature of quantum mechanics, there are many other ways in which noise gets onto a beam of light, particularly laser light. Most of these noise sources don't stem from fundamental laws of physics, like quantum mechanics, and are lumped under the heading of "technical" noise sources. With sufficient care in the experiment design and engineering, they could, in theory, be removed. In practice, we learn to live with many of these technical noise sources; perhaps we couldn't find the cause, or perhaps removing them is beyond our technical ability or resources. The following is an incomplete list of some of the most common technical noise sources on laser beams.

1. Phase and pointing noise due to mirror vibrations or air currents.
2. Amplitude noise due to slight beam clipping on vibrating components.
3. Parasitic interferometers.

4. Intrinsic noise-generating laser dynamics.

5. Insufficiently controlled laser power supplies or current drivers.

6. Noise due to beam scattering in air or on mirrors.

7. Thermal changes causing laser instability.

8. A portion of the laser beam is retroreflected back into the laser cavity causing instability.

9. RF pickup and cross-talk getting amplified and sent to optical modulators.

... et cetera.

Some of these are fairly self-explanatory and it's reasonably obvious how to deal with them if they arise. Others, like parasitic interferometers, scattering, and the intrinsic dynamics of lasers warrant some discussion.

9.3.1 Parasitic Interferometers and Scattering

When a small portion of a laser beam gets back-reflected on a transmissive optic and then re-reflected forward on another optic, we have a situation where two beams of the same wavelength are traveling forward along the optics chain together. The doubly reflected beam is likely to be very weak and only a tiny portion of it will be in the same mode as the main beam. However, that portion is not zero. I'll call the portion of the doubly reflected beam that's in the same optical mode as the main beam the "parasitic" beam. This parasitic beam will interfere with the main beam. If the pathlength taken by the parasitic beam is changing by more than $\lambda/2$, relative to the main beam, then the optical power at the photodetector will oscillate as the interference sweeps through fringes. The corresponding fractional change in the optical power at the photodetector (see Exercise 9.10) is

$$\frac{\Delta P}{P} = 4 \sqrt{\frac{P_{\text{para}}}{P}}, \qquad (9.21)$$

where ΔP is the peak-to-peak power change as the parasitic beam sweeps through fringes, P is the power in the main beam, P_{para} is the power in the parasitic beam. The PSD of the fractional power change is the relative intensity noise $S_{\text{RIN}}(f)$. See Example 9.2, where the RIN due to a realistic parasitic interferometer is presented. The significant contribution to the RIN is one reason it's a very good idea to use transmissive optics with antireflective coatings. If that's not possible, or if the noise due to parasitic interference is too high despite antireflective coatings, you can try the usual trick of rotating transmissive optics slightly so that they're not perfectly normal to the optics chain. That way reflections leave the system and are less likely to have components in the mode of the main beam.

> **Example 9.2 Parasitic Interference versus Shot Noise in an Audio-Frequency System**   Let's consider the case of using a $P = 5$ mW, shot-noise limited, $\lambda = 532$ nm, laser to measure some weak broadband signal that occurs in the audio band. Our photodetector should be sensitive to signals between a few Hertz and something like ten kilohertz – a bandwidth $B \approx 10$ kHz. (The Michelson interferometer

you may have built in Section 5.3 is very likely to have been just such a system even if it wasn't shot noise limited). The relative intensity noise due to shot noise is given by Eq. (9.9). With our numbers

$$S_{\text{RIN\_shot}} = 1.5 \times 10^{-16} \text{ Hz}^{-1}.$$

The RMS level of the relative power fluctuation due to shot noise is then

$$\frac{\Delta P_{\text{RMS\_shot}}}{P} = \sqrt{S_{\text{RIN\_shot}} B}$$
$$= 1.2 \times 10^{-6}.$$

In other words, if you look at the output of the photodetector on an oscilloscope, it would show a trace with a width due to shot noise that is about one millionth of the DC level. If you have enough gain in the signal chain that the DC level is perhaps 10,000 Volts (which would be high-pass filtered out due to pre-gain AC coupling) you would be able to see the shot noise fluctuations on an oscilloscope on the 10 mV scale.

Now consider a parasitic beam formed from 1% of a double retroreflection at un-coated glass optics, each with 4% reflectance. The relative power in the parasitic beam is then $\frac{P_{\text{para}}}{P} = 0.01 \times 0.04^2 = 16 \times 10^{-6}$. If the phase of the parasitic beam is indeed wrapping due to vibrations, air currents, and so forth, then Eq. (9.21) gives

$$\frac{\Delta P}{P} = 1.6 \times 10^{-2}.$$

Parasitic interference will clearly dominate the time-domain data and is likely to dominate in the frequency domain as well since the fringe wrapping rate is likely to be variable. The following figure shows the noise generated by the level of parasitic interference assumed in this example. The motion of the optic was chosen to be a slow semi-random motion on the micron scale that changes direction several times a second, suggestive of buffeting of an optic by an air current.

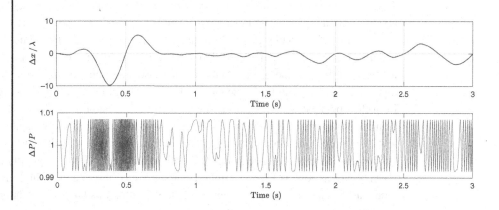

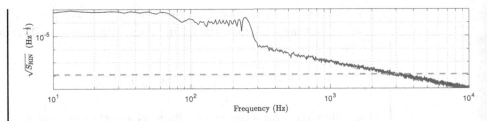

The top panel shows the motion of the retroreflecting transmissive optics involved in generating the parasitic beam. The center panel shows the resulting relative power fluctuations at the output. There is significant fringe wrapping. The bottom panel shows the square root of the power spectral density of the relative intensity. (We've chosen to plot the square root of the PSD to compress the y-axis. This is common practice and the result is called the "amplitude spectral density.") The "cliff" at about 300 Hz is characteristic for small amplitude fringe wrapping. The cutoff frequency is set by the maximum speed with which the fringes wrap due to the optic motion. The dashed line shows the shot noise level, which in this example is completely swamped by parasitic interferometer noise in most of the band. This would not be an acceptable situation for low-noise measurements.

Scattering noise is due to forward-scatter, or reflected back-scatter, of optical field components by dust or optic imperfections. As in parasitic interferometer noise, some of the scattered field will be in the mode of the output beam and able to interfere with it. The main distinction between scattering and parasitic interferometer noise is that scattering doesn't tend to be as bad as parasitic interferometer noise but is harder to get rid of. Also, unless the scattering centers are attached to an optic, the motion they undergo is probably slower than that of a vibrating optic. Scattering noise may therefore enter at lower frequencies than parasitic interferometers.

9.3.2 Intrinsic, Noise-generating, Laser Dynamics

Sometimes the laser output intensity will oscillate for no apparent reason. It may settle down after a while or it may oscillate intermittently. One class of such oscillations are "relaxation oscillations" of the laser field. Some lasers, especially solid-state lasers, have the property that if the pump is suddenly turned off, light will exit the cavity faster than the upper lasing level can be depleted. In other words, the lifetime of the upper lasing level is longer than the lifetime of the cavity. If such lasers are pumped to above threshold, then light will quickly build up in the cavity. The associated increase in stimulated transitions from the upper lasing level causes the population inversion to be depleted more quickly than the pump can make up for. So, the gain will fall below threshold again and the light level will drop. Once that happens, the cycle can begin again as the pump brings the inversion above threshold once more. The oscillation frequency varies greatly between laser types, being as low as several kilohertz in solid-state lasers and as high as gigahertz in diode lasers. Gas lasers on the other hand, don't exhibit this kind of oscillatory behavior.

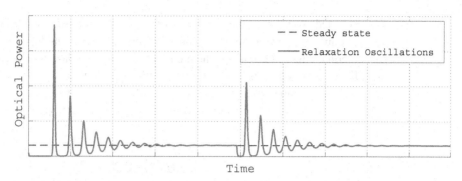

Figure 9.4 A numerical solution to Eqs. (9.22) and (9.23) showing large relaxation oscillations of a laser (also called "spiking"). Depending on the laser, such oscillations may be constantly excited at a low level by small glitches in the input power or other changes in operating conditions.

The behavior can be described by rate equations like Eq. (6.16) for the number of atoms N_2 excited to the upper lasing level and a coupled equation for the photon flux in the laser cavity. For example,

$$\frac{dN_2}{dt} = R_p - \gamma_{21}N_2 - \sigma_{21}N_2\mathcal{N}, \tag{9.22}$$

$$\frac{d\mathcal{N}}{dt} = \frac{\sigma_{21}c}{V}N_2\mathcal{N} - \gamma_c\mathcal{N}, \tag{9.23}$$

where N_2 is the number of atoms in the upper laser state, and $\mathcal{N}$ is the photon flux[4] in the laser cavity, $\mathcal{N} = \frac{I(t)}{h\nu}$. R_p is the pump rate, γ_{21} is the probability of spontaneous transition $2 \rightarrow 1$ of a single atom per unit time, σ_{21} is the cross section for stimulated transition $2 \rightarrow 1$, V is the volume of the optical mode within the gain medium, c is the speed of light, and γ_c is the rate at which the photon flux leaves the cavity through the input or end mirrors or through loss mechanisms. For certain ranges of coefficient values, these equations exhibit oscillatory behavior as shown by their numerical solution in Figure 9.4. The figure shows the optical power (proportional to $\mathcal{N}$) as a function of time. The behavior seen is a set of spiky damped oscillations that settle down to the steady state value given by Eq. (6.18). If the system is disturbed again after settling down, the oscillations can start again. This is also illustrated in the simulation shown in Figure 9.4. About halfway through the time series, the cavity storage time was briefly reduced. This could simulate a bumped laser mirror. A similar effect happens when the pump power is briefly reduced. So, we see that the fundamental rate equations for lasers contain within them the potential for ongoing oscillation in response to environmental disturbances, imperfect power delivery, and so forth.

Another source of intrinsic optical noise in lasers is due to the fact that they can have several spatial modes (TEM$_{mn}$) and/or several longitudinal modes (different number of half-wavelengths within the cavity) excited simultaneously. The different modes may have slightly different wavelengths but all lie within the band of the laser transition and so all

---

[4] The photon flux, $\mathcal{N}$, is the number of photons per unit cross-sectional area, per unit time.

can be excited. Sometimes, the ratio of modes doesn't stay fixed in the laser and this leads to output amplitude variation as the modes compete for photons from the gain medium. Also, thermal effects may change the mode composition of a laser; for this reason it's best to let lasers warm up for a while before use. It's possible to avoid this source of noise by using only single-mode lasers.[5] They are generally lower noise and better behaved, but also significantly more expensive than multi-mode lasers of the same basic type.

Exercises

9.1   Show that the average photon rate $\langle n \rangle$ in a beam with power $P = 50$ W and wavelength $\lambda = 1.064$ μm is about 2.5×10^{20} photons per second.

9.2   Estimate the total number of photons, n, exiting a 1 mW green laser pointer of wavelength 532 nm during the course of a typical hour-long colloquium talk if the laser pointer is used with a 25% duty cycle.

9.3   Write down the first five coefficients a_n in a coherent state, $|\alpha> = \sum_n a_n |n>$ when $\alpha = \sqrt{2}$. What's the expectation value $\langle n \rangle$ for the number of photons in $|\alpha>$? What's the expectation value for the number of photons in a coherent state $|\alpha>$ with $\alpha = \sqrt{2} + i\sqrt{2}$?

9.4   The Poisson probability density $p(n)$ is given by Eq. (9.4) and the corresponding Gaussian probability density is given by

$$g(n) = \frac{1}{\sqrt{2\pi\sigma^2}} e^{-\frac{(n-\langle n \rangle)^2}{2\sigma^2}},$$

where $\sigma^2 = \langle n \rangle$. The proof that these probability densities tend to one another for large $\langle n \rangle$ is not easy. However, with the help of Stirling's formula it's not too difficult to show that the *peaks* of the distributions have the same height as $\langle n \rangle \to \infty$. Show this.

9.5   (Computer problem) Show that adding a sine wave to Gaussian noise modifies the probability density in such a way as to broaden it, and if the RMS amplitude of the sine wave is significantly larger than the RMS amplitude of the noise, the distribution becomes bimodal obtaining two peaks. Explain why that is. You may find it helpful to consider a child on a swing or a mass on a spring as examples of approximately sinusoidal motion.

---

[5] Pure single-mode lasers have only one longitudinal mode present and that longitudinal mode is in a single spatial mode, usually a TEM_{00}. Since a single spatial mode is fairly easy to achieve, and is generally expected from a good quality laser, single-mode lasers also go by the name "single longitudinal mode" or "single-frequency" lasers.

9.6 (Computer problem) *Numerically* integrate a Gaussian probability density, $g(x)$ from $x = -\infty$ to $x = a$ producing $G(a)$, the cumulative distribution function (CDF). Do this for 100 values of a in the range $-4\sigma < a < 4\sigma$. Plot $G(a)$. Check your numerical result by comparing it to the analytic CDF, which you may look up.

9.7 (Computer problem) Write a simple program to calculate the power spectrum of $z(t)$ using fast Fourier transforms, "FFT's." In other words, use $S = |\tilde{z}(f)|^2$ where $\tilde{z}(f)$ is the Fourier transform of $z(t)$. In Matlab the FFT command is "fft" and in Python it can be found in the NumPy package as "np.fft.fft." Enhance your program by adding the ability to calculate an averaged power spectrum. *Hint*: To calculate an average spectrum with n averages, split your time-domain data $z(t)$ into n equal parts. Calculate the power spectra of the individual parts and average them. Test your function on a sine wave and on random noise.

9.8 Estimate the (single-sided) relative intensity noise due to shot noise from a flashlight received by a photodiode if the DC photo-current is 0.1 mA. As with other PSDs, the single-sided RIN is twice the double-sided RIN (where negative frequencies are treated as separate from positive frequencies). What would the DC current need to be for the double-sided RIN to be 1 Hz^{-1}?

9.9 Estimate the RMS level of the signal on an oscilloscope due to shot noise at the output of a Michelson interferometer. Assume the oscilloscope is connected directly to a photodiode at the antisymmetric port of a simple Michelson interferometer operating at mid-fringe. (Connecting the photodiode directly to the oscilloscope is not actually a good idea in practice for reasons explained in Section 2.2.6.) Assume the laser is shot noise limited, has wavelength $\lambda = 633$ nm, and power $P_{in} = 5$ mW. The oscilloscope has an input resistance of 1 MΩ. The capacitance of a 1 mm^2 silicon photodiode in conjunction with the 1 MΩ, AC-coupled oscilloscope acts like an input filter with passband 0.3 Hz−15 kHz. Is this RMS voltage signal visible on a typical oscilloscope? *Hint*: First find the average (DC) photocurrent generated by the photodiode. Then find the corresponding DC voltage by multiplying by 1 MΩ. You can assume the oscilloscope doesn't affect the photocurrent in any way except that it is insensitive to fluctuations outside the aforementioned passband.

9.10 Show that Eq. (9.21) is correct. *Hint*: Consider the case where the parasitic beam and the main beam interfere constructively and compare it with the case where they interfere destructively. Recall that $I \propto |E|^2$.

9.11 (Computer problem) Reproduce the center panel in Example 9.2, except allow the optic motion to be sinusoidal with frequency 2 Hz and amplitude 2 microns. (Assume you have a parasitic interferometer generated between two optics without antireflective coatings that reflect 4% of the incident power. Assume that 1% of the forward-traveling, doubly reflected light is in the same optical mode as the main beam. Assume a 5 mW noiseless laser with wavelength 532 nm.) Then repeat the exercise but with antireflective coatings on the optics that reflect only 0.25% of the incoming light power. How much is the output power fluctuation reduced by the antireflective coatings? Draw the time-domain traces for each case on the same graph.

9.12 (Computer problem) Solve Eqs. (9.22) and (9.23) together numerically using a built-in ordinary differential equation solver with either Matlab or Python. Without worrying too much about the "correct" values of the parameters, see if you can get oscillatory behavior for any values of the parameters. *Hint*: Start off with all coefficients of N_2 and $\mathcal{N}$ equal to one except set $\gamma_{21} = 0$. Start with $R_p = 1$ also. Let t run from 0 to 100 in these units. Once you have the code working, try adjusting the parameters until you find oscillations. Make sure your initial condition for $\mathcal{N}$ is not $\mathcal{N}(t = 0) = 0$ or $\frac{d\mathcal{N}}{dt}$ will remain zero.

9.4 Experiment: Shot Noise

Objectives

1 Investigate optical noise in the time and frequency domains.
2 Interpret power spectra.
3 Characterize the general properties of optical noise.
4 Investigate the dependence of shot noise on optical power.

Equipment Needed

- A photodiode with active area about 3–10 millimeters across.
- A *low noise* photodiode amplifier. Transimpedance in the vicinity of 1–10 $k\Omega$ works well. If a low noise amplifier with variable transimpedance is available, it can be used. The amplifier may also be part of a single package that includes the photodiode.
- A flashlight with an incandescent lamp.
- A flashlight with an LED lamp.
- A 2–5 mW helium-neon laser.
- Several neutral density filters of modest optical density (e.g., OD: 0.2, 0.4, 0.6, 1.0). Filters can be combined. You'll need enough filters to get *at least* five different transmittance levels.
- An oscilloscope.
- A low noise, audio-band, spectrum analyzer covering at least 100 Hz to 20 kHz range (e.g., the classic SR780). You can also use a data acquisition system and make the spectra in real time or off-line. This is more versatile but requires care to do correctly.[6]
- Optional: Short focal length (< 200 mm) lens to collect the light from the flashlights. The larger the lens, the more light can be collected, so I recommend a lens with 5 cm diameter or more. (This may not be needed if the flashlights are bright enough.)

Optical Noise

Every source of light has noise fluctuations in amplitude and phase. These noise types go by the names: "amplitude noise" and "frequency noise" since the time derivative of phase is frequency. Measuring frequency noise is somewhat tricky because it requires using some sort of optical frequency discriminator. A stable optical cavity (see Chapter 7) can be used, but its setup is too involved for this lab. Amplitude noise on the other hand, can be measured directly with a photodiode. You will investigate the characteristics of amplitude noise on the three main classes of light sources: incandescent lamp, LED, and laser. Your photodiode should be sensitive enough to discern "shot noise," which arises from quantized light energy (photons) registered by a photodiode. Like the sound of rain, shot noise is broadband.

Shine the flashlight with the incandescent lamp at a photodiode. You may need to put a lens in front of the photodiode in order to focus as much light as possible onto the active

---

[6] Make sure you have DC blockers and very low noise amplifiers immediately after the photodiode. You may also need anti-alias filtering and whitening before the signal is acquired.

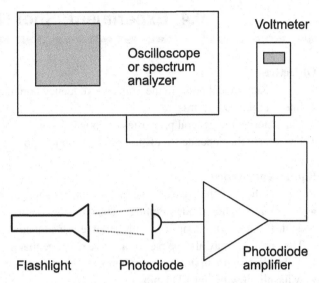

Figure 9.5 The setup for measuring shot noise. The DC Voltage shown by the voltmeter is proportional to the total optical power absorbed by the photodiode.

area of the photodiode. Make sure no other light is reaching the photodiode. Figure 9.5 gives an overview of the setup.

> *Recommendations*: Either turn down the room lights all the way or put a cardboard box over the photodiode and cut out a hole so that only light from the flashlight reaches the active area of the photodiode. A simple tube can serve the same purpose. Make sure the output of your photodiode amplifier is not close to saturation. Find the saturation level by shining a very bright light onto the photodiode. It is saturated when the output level doesn't respond to small changes in the light level. Now, using the flashlight, adjust the amount of light reaching the photodiode (and/or adjust the amplifier gain) so that the output level is between 25% and 75% of the saturation level.

First, connect the photodiode amplifier output to an oscilloscope. Set the oscilloscope input to be AC coupled and increase the voltage sensitivity until you see the noise (probably in the milliVolts range). What can you say about the noise as a function of *time*? Is it random? Are the features generally the same over time ("stationary") or do they fluctuate? Measure the RMS amplitude of the noise.

Replace the oscilloscope with the spectrum analyzer. Ask your instructor to help you set up the spectrum analyzer if you have not used one before. It should be in the lowest noise configuration possible. It must be AC coupled and have the largest possible amount of input amplification (lowest input range) while still accommodating your signal without saturation. Adjust the spectrum analyzer so it displays the band between about 100 Hz and 10 kHz and set it to average the incoming data for a few seconds. What can you say about the structure of the noise in this band as a function of *frequency*? If there are features, can you identify their source? For example, noise due to the AC power in the lab will appear as sharp peaks at multiples of the line frequency (50 Hz or 60 Hz, depending on your region). Shot noise on the other hand is supposed to be white (flat spectrum). Are there any frequency stretches within this band where the noise is white?

You'll now vary the amount of light incident on the photodiode using neutral density filters. At each light level, record the power spectral density S_V of the broadband noise in units of V^2/Hz at two different frequencies f_1 and f_2 of your choice in the kHz regime. (For example, $f_1 \approx 2$ kHz and $f_2 \approx 8$ kHz tend to be good choices.) Some analyzers have a choice between peak-to-peak units (V_{pp}^2/Hz) and root mean square units (V_{rms}^2/Hz). Use, RMS units. Make sure that the frequencies you choose are not contaminated with sources of narrowband noise. In other words, avoid sharp peaks in the power spectrum. At each light level, also record the DC voltage, V_{DC}, from the photodiode amplifier as measured on the voltmeter. At the end, block the light entirely and measure the dark noise level. Plot $S_V(f_1)$ and $S_V(f_2)$ versus V_{DC}.

The DC voltage level coming from the photodiode is proportional to the optical power. So, Equation (9.8) predicts a linear relationship between the broadband power spectrum amplitude and the DC voltage. As explained earlier in the chapter, the prediction of a linear relationship ultimately relies on the picture that light is quantized. Do a linear fit to $S_V(f_1)$ versus V_{DC} and do a second linear fit to $S_V(f_2)$ versus V_{DC}.

According to Eq. (9.11), we can also write the PSD of the photocurrent due to shot noise as

$$S_i(f) = 2ei. \tag{9.24}$$

Multiplying both sides by R^2 and using V=iR on the right-hand side gives the PSD of the voltage presented to the spectrum analyzer

$$S_V(f) = 2eRV, \tag{9.25}$$

where V is the average (or DC) voltage. Do your S_V vs. V_{DC} fits agree *quantitatively* with this equation after subtracting the dark noise level?

> *Recommendations*: Make sure you understand the units on the spectrum analyzer. If your spectrum analyzer has the ability to display physical units like V^2/Hz use them in preference to any sort of decibel unit. (Decibel units are logarithmic and will need to be converted to $\frac{\text{Volts}^2}{\text{Hz}}$.) Unless it's flat, you will need a transfer function for the photodiode amplifier. (You will need the response at f_1 and f_2 compared to the response at DC.)

Repeat this measurement with a HeNe laser as the light source. At frequencies below a few kilohertz, a typical HeNe will be limited by technical sources of noise on the light. The broadband PSD should therefore go like the square of the DC voltage. Is this the case? To compare with the flashlight behavior, plot $S_V(f_1)$ and $S_V(f_2)$ versus V_{DC} as before on the same graph as the flashlight data. If you find that the HeNe is good enough to be shot noise limited at the frequencies you first investigated, try finding a particularly noisy frequency and show that its PSD does in fact go like V_{DC}^2 at those frequencies.

Investigate the noise behavior of the light from the LED flashlight in the same way. Does its noise resemble the laser's or the incandescent lamp's?

Ideas for Further Investigation

The response of a photodiode amplifier usually decreases with frequency. Use the transfer function of the amplifier (supplied with the amplifier or by your instructor) to correct your spectra for this. Are the spectra white after correcting for this? If not, why not?

The reason we used neutral density filters in this experiment was to ensure that the spectrum of the light didn't change along with the irradiance. If we had instead changed the voltage going to the incandescent lamp, a reduction in the voltage would lead to a reddening of the output spectrum along with the change in the output irradiance. How would this affect the measurement you made?

Set up a 50/50 beamsplitter and direct equal amounts of light to each of the two identical low noise photodiodes. Endeavor to keep the lengths of the two paths equal. Since shot noise is generated by the detection of photons at each of the photodiodes separately, the shot-noise contribution to the two photocurrents should be uncorrelated. For a shot noise limited source, the cross-spectral density between the two photodiode signals should therefore be smaller than the power spectrum of either of the individual photodiode signals.[7] Verify this and comment on whether this suggests that the noise is indeed generated by wave function collapse. If you have time, compare this behavior to that of a source that is not shot-noise limited.

---

[7] The power spectrum of a single, amplified, photodiode signal voltage should be identical to the cross-spectral density of the same signal with itself. You can check this directly by connecting two ports of the spectrum analyzer or data acquisition system to the same signal using a simple "T" (not a power splitter). Compare the power spectrum of either port to the cross spectrum between the ports.

Appendix A Analyzing and Displaying Data with Matlab and Python

This chapter is intended as a place to start for beginning programmers with little or no experience with Matlab and Python.[1] The idea is to give a basic introduction to Matlab and/or Python that enables you to analyze and display data quickly. After this introduction, you should have a feeling for how the language works and be able to complete basic data tasks. Hopefully, you'll be motivated to pick up additional material and learn the language of your choice in some depth. Do heaps of Internet searching while coding and try to understand and remember what you find so as to avoid the same search next time. (Try not to just cut, paste, and forget.) The answer is usually out there and there's almost certainly someone who has already been in your shoes, and someone who has taken the time to help. Make an effort to remember the syntax on your own; it will make coding quicker and far more enjoyable.

A.1 Matlab versus Python

We'll look at two of the most popular programming languages in science: Matlab and Python, side by side. This will give you an idea of the similarities and of the crucial differences. If you already know either Matlab or Python very well, you can take this as an opportunity to learn the other. If you know neither Matlab nor Python, concentrate on learning one of them well but keep an eye on how things are done in the other language. Do not try to learn both Matlab and Python at the same time. Choose one! My advice is to first learn whichever language is being used by your friends and colleagues.

Matlab was designed as a scientific programming language, a "Matrix Laboratory" with roots from Fortran starting in the late 1970s and maturing in the 1990s. Matlab's great strength is that you don't have to worry about fancy programming concepts. You just have to understand its basic syntax and especially how to deal with matrixes and arrays. The simplicity with which it handles matrixes and arrays makes it a joy to use with data. Matlab is not free although students are offered the software at a greatly reduced price. Matlab can also be extended by purchasing additional components from Mathworks, the producer or Matlab.

Python is a newer language, first designed in the late 1980s and maturing in the 2000s. Python's strength is that it is a general-purpose programming language that is designed to be highly extensible via add-on packages. It's also free, as are most of the packages. Python is therefore very flexible in what it can do and it's easy for users to obtain the

---

[1] Code was written in Matlab 2018b and Python 3.7. The code in this chapter and the necessary accompanying material are available at https://github.com/CambridgeUniversityPress/FirstCourseLaboratoryOptics.

tools they need. You add on the packages you need to do the tasks you want. The core scientific add-on packages, NumPy and SciPy, give Python approximately the same core capabilities as Matlab. Since there are thousands of add-on packages available for Python, chances are that something exists to address most specialized needs. Examples include: astronomy (astropy), molecular biology (biopython), meteorology (metpy), general relativity (einsteinpy), symbolic math (sympy), and so on. This extensibility, and the fact that Python is designed as a general-purpose programming language, means that getting started in scientific programming with Python may be slightly more complicated than getting started with Matlab.[2]

A.2 Basic Coding

Here is code that plots a sine wave.

MATLAB

```
theta = -4*pi:0.1:4*pi;
plot( theta, sin(theta) );
shg;
```

PYTHON

```
import numpy as np
import matplotlib.pyplot as plt

theta = np.arange(-4*np.pi,4*np.pi,0.1)
plt.plot( theta, np.sin(theta) )
plt.show()
```

In the Matlab version, we first define a long list of numbers between -4π and 4π separated by 0.1 and assign them to the variable theta. (Such a list is known as a "vector" in Matlab. It's really just an array consisting of a single row or single column.) Then we plot that vector of numbers against the sine of its elements. The final line stands for "show graph" and just brings the plot window to the front. As you can see, the Python version has more to it. The third and fourth line are basically the same as the Matlab version but with "np" and "plt" appearing in places. Also, the way to define a vector of numbers is a little different in Python. But there are two extra lines at the beginning of the Python code. Due to the fact that Python is designed to be extended, the first thing we have to do is bring in the extension package (NumPy) that defines the sine function and the irrational number π. NumPy also allows us to make vectors via the "arange" method. The next thing is to bring in the extension package that allows plotting (matplotlib.pyplot). We also choose to give the extension packages nicknames, np and plt respectively, by which the package contents can be referenced later. So, for example, plt.plot is the plotting function we need and np.sin is the sine function.

If you're new to Matlab or Python, you should try running this code. If you are using Matlab, the integrated development environment (IDE) opens by default. If you are using Anaconda Python, you should start the "Spyder" IDE. (Other IDE's exist but I'll assume for the purposes of this tutorial that you're using Spyder.) In the editor window, type in the code above. Then press the green "play" button at the top of the window. This saves and runs the code and works the same way in Matlab's IDE and in Spyder. (If the file hasn't been saved before, the IDE may prompt you for a name. You can save it under any name you like.)

---

[2] To obtain a full system including an integrated development environment called "spyder" I recommend downloading "Anaconda Python." At the time of writing, it is free.

You've probably noticed that there is a console window in both Matlab's IDE and in Spyder that (among other things) shows the name of the file you've just run. In fact, the console is where the program actually launches. When you pressed the green arrow, the IDE launched your program just as you could by hand, by entering the appropriate command into the console. You can enter any command directly into this console window and the command interpreter will try to execute it. You can also use it as a calculator since it also understands arithmetic operations. For example, try entering

MATLAB

```
>> 2+2
```

PYTHON

```
In [1]: 2+2
```

Don't enter the ">>" part (Matlab) or the "In [1]:" part (Python). Those are just the console prompts. I included them to emphasize that these commands are to be entered directly into the console and not into the editor. When you press enter, the output should show the result of adding two and two. You can also try other things like

MATLAB

```
>> sin(pi/4)
```

PYTHON

```
In [1]: import numpy as np
In [2]: np.sin(np.pi/4)
```

By the way, when a Python program (written in an editor) is run, nothing is output to the console window without an explicit print command. In Matlab, it's the other way round; the output of any statement in the program is displayed in the console by default. To suppress this, we need to put a semicolon at the end of each Matlab command.

The code below shows how to define arrays directly. You can then extract their contents, transpose them, do matrix arithmetic with them, and more. Since this code is to be entered directly into the console window, you will see the result of each command displayed as it is entered.

MATLAB

```
>> A = [ [2,0,1]; [0,1,3]; [-1,1,0] ]
>> B = [ [1,1,2]; [1,2,-2]; [1,4,1] ]
>> A(2,1)        % single element
>> A(2,:)        % entire row
>> A(:,2)        % entire column
>> A.'           % transpose
>> A*B           % matrix mult.
>> A.*B          % element-wise mult.
```

PYTHON

```
In [1]: import numpy as np
In [2]: A = np.array([ [2,0,1], [0,1,3], [-1,1,0] ])
In [3]: B = np.array([ [1,1,2], [1,2,-2], [1,4,1] ])
In [4]: A[1,0]        # single element
In [5]: A[1,:]        # entire row
In [6]: A[:,1]        # entire column
In [7]: A.T           # transpose
In [8]: A @ B         # matrix multiplication
In [10]: A * B        # element-wise mult.
```

The expressions starting A = and B = show you how to define arrays directly. Basically, each row is given by a tuple like [2,0,1]. One subtle difference between Matlab and Python's way of doing things is that in Python the rows are separated by commas whereas in Matlab they need to be separated by semicolons. The three subsequent expressions are for indexing into an array, that is for extracting a subpart of the array. The right-hand side of each of these lines contains a comment, indicated by the comment character "%" or "#"; the remainder of the line following the comment character is ignored by the command interpreter. Note that Python array indexing starts at 0 while Matlab's starts at 1, so Matlab's row and column numbers are always one greater than the corresponding Python row and column numbers. A(2,1) in Matlab means the element in the second row, first column. In Python, the same element is accessed as `A[1,0]`. Also note that Python uses square brackets while Matlab uses round brackets for indexing into arrays. In the next two expressions, the colon means "all elements." So, they return the entire second row of `A` and the entire second column, respectively. The next line shows how to take the transpose of a matrix. Then we see two possible ways of multiplying matrixes: using the rules of matrix multiplication (you should see that $AB \neq BA$) and elementwise multiplication where corresponding elements are simply multiplied together.

Data can be entered manually into an array via an editor. For example, the results of a measurement might be entered and displayed as follows. Note the use of the comment character to label the columns of the data. The resulting plots are illustrated following the code.

MATLAB

```
data = [
%    X      deltaX   Y       deltaY
     0.9    0.2     2.1      0.2
     1.4    0.3     3.4      0.3
     2.1    0.2     3.1      0.6
     2.7    0.2     4.8      0.5
     3.4    0.2     5.1      0.2
     ];

dxneg = data(:,2);      % left x error bars
dxpos = data(:,2);      % right x error bars
dyneg = data(:,4);      % lower y error bars
dypos = data(:,4);      % upper y error bars

errorbar(data(:,1),...  % cont. on next line
         data(:,3),dyneg,dypos,dxneg,dxpos,'o');
axis([0 4 0 7]);        % set axes limits
grid('on');

hold('on');             % allow overplotting
plot([0.5,3.75],[1.5, 5.8],...
     'r-');             % "by eye" fit

xlabel('{\Delta}L   ( nm )');
ylabel('T   ( {^\circ}C )');
shg;
```

PYTHON

```
import numpy as np
import matplotlib.pyplot as plt

data = np.array(# \=line continnuation char.\
[ #  X       deltaX   Y       deltaY \
  [  0.9,    0.2,    2.1,     0.2 ],\
  [  1.4,    0.3,    3.4,     0.3 ],\
  [  2.1,    0.2,    3.1,     0.6 ],\
  [  2.7,    0.2,    4.8,     0.5 ],\
  [  3.4,    0.2,    5.1,     0.2 ]\
])

deltaX = data[:,1]
deltaY = data[:,3]

plt.errorbar(data[:,0],data[:,2],\
    xerr=deltaX,yerr=deltaY,fmt='o')
plt.axis([0,4,0,7])
plt.grid(axis='both')

# Python default is to allow overplotting
plt.plot([0.5,3.75],[1.5, 5.8],\
     'r-');             # "by eye" fit

plt.xlabel('$\Delta L$   ( nm )')
plt.ylabel('$T$   ( $^\circ C$ )')
plt.show()
```

Manual direct data entry in Python is a bit clunkier than in Matlab but only slightly. Both systems produce very similar results. However, data isn't often entered manually unless the number of data points is under a few dozen. Often, data will be written to disk directly from the instruments, perhaps as binary-encoded numbers, text-encoded numbers, possibly as an image, and so forth. An important part of using electronic data gathering systems is the ability to import all the different data formats produced by our instruments. We can't cover all possibilities but shall concentrate on importing digital camera images because much of our data is captured with cameras.

The following code shows how to import camera data and plot the result as an intensity map of the image

MATLAB

```
A = imread('myphoto.jpg');
A = double(A);
A=mean(A,3); %only do if image was color
pcolor(A);
shading('flat'); colormap('bone');
```

PYTHON

```
import numpy as np
import matplotlib.pyplot as plt
import matplotlib.image as mpimg

A = mpimg.imread('myphoto.jpg');

A=np.mean(A,2) #only do if image was color
plt.pcolormesh(A,\
        shading='flat',cmap='bone');
```

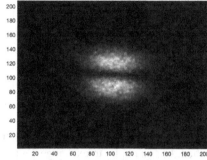

The line containing "imread" is the one that actually reads in the picture. (In Matlab, images are read in as unsigned short integers, 'uint8,' and should be converted to double precision floating point numbers, 'double,' before analysis. That's accomplished by the line A = double(A);.) After reading the image, we convert it to total pixel intensity values by taking the mean of the three colors. The image is read in as a $M \times N \times 3$ array, so after taking the mean over the third array dimension (known as array 'pages'), we're left with a 2D, $M \times N$ array where each pixel value is proportional to the optical power it receives. Finally, we plot an intensity map of the array that is shown. It's a black and white version of the original image – in this case a TEM_{01} mode from a HeNe laser cavity. The x-axis and y-axis values are pixel numbers. Code in Appendix B.1 shows how to display a cut through the image (single row or column) and how to smooth the resulting curve, which is quite noisy due to laser speckle.

A.3 Loops and Branching

Branching structures allow you to choose between execution of different possibilities based on different conditions. The most used branching structure is the "if" structure. In its most basic version, it looks like the following.

MATLAB	PYTHON

```
a = 10;
b = 3;

if a/b==3
    disp(' ')
    disp('Matlab is doing integer
    arithmetic.')
else
    disp(' ')
    disp('Matlab is doing floating point
    arithmetic.')
end
```

```
a = 10;
b = 3;

if a/b==3:
    print(' ')
    print('Python is doing integer
    arithmetic.')
else:
    print(' ')
    print('Python is doing floating point
    aritmetic')
```

The expression following the keyword if must evaluate to True or False. Such an expression is called a "conditional expression." The == sign is the *comparison* operator for equality. (The single = is the *assignment* operator, used to assign values to variables as in the first two lines of code.) Other common comparison operators are <, >, <=, =>, and ~= (Matlab), != (Python). The negation operator is ~ in Matlab and ! in Python. In both Matlab and Python, logical AND and logical OR are & and | respectively. If the conditional expression in the above-mentioned code evaluates to True then the indented code block immediately following it is executed. If there is also an else statement as shown above, it gets executed when the prior conditional expressions evaluate to false. (It is also possible to check additional conditional expressions using elseif/elif statements before the final else.) Please read up on the help documentation on the if statement as it's one of the fundamental programming statements!

Loops are the other fundamental programming structure. They are used to repeat a block of code until some condition is satisfied. For example, to exchange every other entry in a vector, you could write something like

MATLAB	PYTHON

```

                                   import numpy as np

vec = (1:9);                       vec = np.arange(1,10)
s = 2;                             s = 1
while s <= length(vec)             while s <= len(vec)-1:
    tmp = vec(s);                      tmp = vec[s]
    vec(s) = vec(s-1);                 vec[s] = vec[s-1]
    vec(s-1) = tmp;                    vec[s-1] = tmp
    s = s + 2;                         s = s + 2
end
disp(vec)                          print(vec)
```

First the vector vec is defined. It's just [1,2,...,9]. (In Python the upper number in the range definition is excluded, in Matlab it's included.) The commands length(vec) (Matlab) and len(vec) (Python) return the length of the vector vec, giving 9. So, the while loop will execute while the counter runs 1–9 (Matlab) or 0–8 (Python). While loops can be used to iterate code a predetermined number of times, as we have done here, but they can also contain conditions that become true only after an undetermined number of iterations, or perhaps even never (in which case we have an "infinite loop"). Since iterating a block of code a predetermined number of times is such a common task, there's a shortcut for it known as a for loop. The equivalent code to the above-mentioned code but using a for loop would look like

MATLAB

PYTHON

```
import numpy as np
```

```matlab
vec = (1:9);
for s=2:2:length(vec)
    tmp = vec(s);
    vec(s) = vec(s-1);
    vec(s-1) = tmp;
end
disp(vec)
```

```python
vec = np.arange(1,10);
for s in range(1,len(vec),2):
    tmp = vec[s]
    vec[s] = vec[s-1]
    vec[s-1] = tmp

print(vec)
```

The initial value of the counter has disappeared as has the counter increment. The expression following the while keyword is a termination condition while the expression following the for keyword could be interpreted as: "The counter should take on all the values of this vector, in turn."

A.4 Functions

In principle, you now have all you need to start coding in Matlab or Python. However, a method of breaking code into reusable parts known as functions, is so useful that it has become central to coding in most languages. A function is a block of code that is given a name and can be called to run from some other code, or from the console. However, we also define a set of "input variables" whose values we send to the function when we call it. The function then does whatever it is designed to do, probably using the values of the input variables we sent for some purpose, then it exits and returns control to whatever code invoked it. As it exits, it may also return the values of some "output variables," which could contain, say, the result of some calculation done by the function. Those output variables are then available to the invoking code. So, this is a way of farming out work, to "divide and conquer" a difficult or complex task.

It's a good idea to keep the functions you write. You are more likely than not to find an opportunity to reuse them. To start writing a function, open a new file in an editor and enter the syntax below. For now, be careful to save your function in the directory from which it will be called. In Matlab functions for general use are kept one to a file and the filename should be the same as the function name. In Python, it's easier to group functions by category, several to a file. The filename should be representative of the category. In this example, the Matlab filename is "myfunction.m" and the Python filename is "myfunctions.py." The syntax for writing a function is as follows.

MATLAB

PYTHON

```matlab
function [c,d] = myfunction(a,b)
    % Returns the sum and difference of
    % its arguments. You can type
    % help(myfunction) in the command
    % window to display this comment
    % block.

    c = a + b;
    d = a - b;

end
```

```python
def myfunction(a,b):
    """Returns the sum and difference of
    its arguments. You can type
    help(myfunction) from the console
    to display this comment block,
    known as the docstring."""

    c = a + b
    d = a - b

    return  c,d
```

To call these functions from the console (or from another piece of code) the files containing the function must be in the current working directory or in the Matlab/Python path. So, for example,

MATLAB   PYTHON

```
                                     In[1]: import myfunctions as mfs
>> [x,y] = myfunction(2,5)           In[2]: x,y = mfs.myfunction(2,5)
```

The output of the function is assigned to the variables x, y. Note that in Python you import the file containing the function before calling the function, just as you would import an external package like NumPy in order to access its functions. (You could also add other functions to this same file and they would be available in the same way.) The arguments you send to your function when you call it don't have to be scalars. As long as the internal function operations are compatible with what you send, the function will accept them. This is an example of dynamic typing, where the interpreter is flexible with the variable types being provided. For example,

MATLAB   PYTHON

```
                                     In[1]: import numpy as np
                                     In[2]: import myfunctions as mfs
>> m = [1,4,7];                      In[3]: m = np.array([1,4,7])
>> n = [3,5,7];                      In[4]: n = np.array([3,5,7])
>> [x,y] = myfunction(m,n);          In[5]: x,y = mfs.myfunction(m,n)
>> disp(x)                           In[6]: print(x)
>> disp(y)                           In[7]: print(y)
```

A.5 Putting It Together

We'll now put together a potentially useful program out of the coding structures covered so far. It will take a folder full of beam images and find the maximum irradiance in these images. The code consists of a main program called max_irrad_vs_z and a subroutine (function) called get_image_max.

MATLAB   PYTHON

Main Program   *Main Program*

```
% Plots the max irradiance vs position from    # Plots the max irradiance vs position from
% a folder of beam images                       # a folder of beam images

                                                import numpy as np
                                                import matplotlib.pyplot as plt
                                                import os
                                                from imageproc import get_image_max

image_folder = 'sample_images';                 image_folder = 'sample_images'
image_extension = '.tif';                        image_extension = '.tif'
nsmooth = 32;                                    nsmooth = 32
```

```
images = dir(image_folder);
posvals = [];
maxvals = [];

for s = 1:length(images)
    [fdir,fname,fext] =
    fileparts(images(s).name);
    fullpath = fullfile( ...
            images(s).folder,
            images(s).name);
    if strcmp(fext,image_extension)
        posvals = [posvals, ...
            str2num(fname)];
        maxvals = [maxvals, ...
            get_image_max(fullpath,
            nsmooth)];
    end
end

plot(posvals,maxvals,'s','linewidth',2);
grid('on');
xlabel('Position');
ylabel('Max. Irradiance');
shg;
```

```
images = os.listdir(image_folder)
posvals = np.array([])
maxvals = np.array([])

for s in range(1,len(images)):
    fdir,fnameext =
    os.path.split(images[s])
    fname,fext = os.path.splitext(fnameext)
    fullpath = os.path.join(image_folder,
    fnameext)
    if fext==image_extension:
        posvals = np.concatenate((posvals,\
            [float(fname)]));
        maxvals =  np.concatenate((maxvals,\
            [get_image_max(fullpath,
            nsmooth)]))

plt.plot(posvals,maxvals,'bs',linewidth=2);
plt.grid(True)
plt.xlabel('Position')
plt.ylabel('Max. Irradiance')
plt.show()
```

Subroutine

Subroutine

```
% Gets the brightest pixel in a smoothed image
% file: image file path, n: # pixels to avg.




function maxval = get_image_max(file,n)
    A = double(imread(file));
    A = mean(A,3);
    A = imgaussfilt(A,round(n/2));
    maxval = max(max(A));
```

```
# Gets the brightest pixel in a smoothed image
# file: image file path, n: # pixels to avg.

import numpy as np
import matplotlib.pyplot as plt
from scipy.ndimage import gaussian_filter

def get_image_max(file,n):
    A = plt.imread(file)
    A = np.mean(A[:,:,0:2],2)
    A = gaussian_filter(A, np.round(n/2))
    maxval = np.max(A)
    return maxval
```

The code can be usefully applied to the case where images are acquired at various locations z along the optic axis of a beam. The camera settings, including exposure, must be kept fixed between images. If the images are taken around a waist (focus), the graph produced can be used to calculate the Rayleigh range and the waist size. See Figure A.1.[3] I've added some annotations to the figure to calculate the Rayleigh range for this particular beam waist.

The code works in the following way. First, the folder containing the images is specified along with the filename extension for the images being used. The next line is the smoothing factor to be used by the function `get_image_max`, which returns the maximum irradiance in an image after "smoothing" it. The next block of code gets the list of files in the image folder and also sets up the (initially empty) vectors for storing the maximum irradiance values ("maxvals") and the optic axis positions z where they occur ("posvals"). The z-position is assumed to be indicated by the name given to the image file by the experimenter and corresponds to the location along the beam where the image was recorded. The next

---

[3] The images used to produce Figure A.1 are available, along with the above-mentioned code, at https://github.com/CambridgeUniversityPress/FirstCourseLaboratoryOptics. These are the same beam images used to produce the image on the cover.

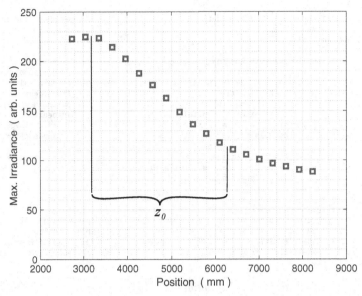

Figure A.1   Output of the code to find the maximum irradiance. The Rayleigh range z_0 is the distance from the waist at which the maximum irradiance has fallen in half. The waist position is at approximately 3200 mm with max irradiance approximately 226 units. The Rayleigh range has been reached when the irradiance has fallen to about 113 units, at about 6250 mm. So the Rayleigh range is about $z \approx 3050$ mm. The wavelength was $\lambda = 633$ *nm*, so Eq. (4.6) mentioned in Chapter 4 leads to a waist size $w_0 = \sqrt{\frac{\lambda z_0}{\pi}} \approx 0.8$ mm.

step, implemented with the for loop, reads each image file and in turn stores its z–position (obtained from the filename) and maximum irradiance value. Immediately after the for statement, there are two or three lines whose function is to get the various parts of the filename of the current image (i.e., folder containing the image, image file name, and the filename extension). These are implemented slightly differently in Matlab versus Python but the outcome is the same. (In Python, we import the os package to accomplish this.) The if statement checks whether the current file has the correct extension to be an image. If so, we get the position of the image from the filename and concatenate it onto the end of the vector posvals. That way, we build up a vector containing all the positions at which the images were taken. Then we get the maximum irradiance value from the image by calling the function get_image_max, which we have written in a separate file. The result is concatenated onto the vector maxvals. Finally, the maximum irradiance values are plotted versus position.

The subroutine get_image_max is a function that takes two arguments: the full file path to an image, and the number of pixels over which to smooth the image. The function reads in the image and converts it to greyscale by taking the mean of all three color layers of the image. The smoothing is done by running over the image in both directions, blurring/smoothing any sharp features by averaging over nsmooth pixels using a Gaussian weighting function. That is accomplished in Matlab with the `imguass` function, which requires the Image Processing toolbox and in Python by `gaussian_filter`, which is available in SciPy. After the image has been smoothed, the `max` function is used to get the maximum pixel value of the smoothed, greyscale image. (In Matlab `max` has to be applied twice because the first time it gives a column vector of the maxima in each row of A.)

Try running the code on the sample images (available online) or with your own images. Just remember that the camera settings should be fixed between images, so it's worth making sure your camera isn't saturated, at the beam waist or any other location, before you start.

The best way to learn how the code works is to change it. Try different values of `nsmooth`. You could also try entirely different ways of smoothing the images by rewriting the `get_image_max` function. You can try to have the code automatically extract the Rayleigh range, z_0, and calculate the waist radius ω_0. If you want a challenge, try adding code to fit the data to the appropriate functional form.

Exercises

A.1 Use the console to find the real and imaginary parts of $e^{i\pi/4}$. (An internet search may help.)

A.2 Use the console to demonstrate that given two 2×2 matrixes, A and B, generally $AB \neq BA$.

A.3 Write a script that reads in a digital photograph, converts it to greyscale, and then plots its "negative" where light hues are shown as dark hues and vice versa. As an optional follow-on challenge, find out how to plot your image as a surface whose height is proportional to the value of a pixel. (Simple images will work best.)

A.4 Measure your reaction time ten times by having a friend drop a ruler between your fingers. (If you've never done this before, do a quick Internet search to see how it's done.) Write a script that plots your reaction time data versus measurement number. Your graph should show the estimated measurement uncertainties using vertical error bars. Also, on your graph plot a dashed horizontal line indicating the average reaction time.

A.5 Write a script that plots the path of a ray of light as it passes through a glass plate of thickness 1 cm. The incident angle of the light ray should be 60° (measured from the plate's surface normal). The light ray is emitted from the origin and the glass plate is perpendicular to the x-axis at $x = 2$ cm. Your code should calculate the ray's path and use Snell's law to calculate the angle changes. Make sure your axes are appropriately labeled.

A.6 Consider the following parametric curve for an epitrocoid. Here, a and b and h are constants and the parameter is ϕ that runs from 0 to 2π.

$$x = (a + b)\cos\phi - h\cos\left(\frac{a+b}{b}\phi\right) \tag{A.1}$$

$$y = (a + b)\sin\phi - h\sin\left(\frac{a+b}{b}\phi\right) \tag{A.2}$$

Write a script that plots the epitrocoid for $a=2$ and $b=h=1$. Then choose a few different values for a, b, and h for which to plot curves and hand in the three you find most interesting. (Note that the parameter ϕ normally runs from 0 to 2π but you can experiment with a larger interval too. You may find that especially interesting for non-integer values of a and b.)

A.7 Write a function that takes an argument x and returns the value of $|\tan x|$ unless the value is larger than 10. In that case, the function should return the string "`too big`."

A.8 Write a function that takes two arguments: a vector x of numbers and a positive integer n. The function should return the product of every n^{th} element in x, starting with the first.

A.9 Write a script that finds the first 100,000 prime numbers. Then produce a histogram of the distance between adjacent prime numbers for the first 100,000 primes that you found. If you're using Matlab, you may not use the "factor" or "primes" functions. The Internet is a good resource though. While testing your program, only try to find the first 100 or 1,000 primes. After you get it working, set it to search for the first 100,000 primes.

A.10 We can define a sequence S_n through the following rules:

A. The first element of the sequence is: $S_1 = d$ where d is a positive integer.

B. The remaining elements are defined via:

$$S_{n+1} = \begin{cases} \frac{S_n}{2} & S_n \text{ even} \\ 3S_{n+1} & S_n \text{ odd} \end{cases}$$

For example, if the first element has the value $d = 3$, the sequence will be $S = 3, 10, 5, 16, 8, 4, 2, 1, 4, 2, 1, \ldots$ where the combination $4, 2, 1$ then gets repeated indefinitely.

It appears that all such sequences eventually reach $S_n = 1$ (and then just repeat 4,2,1 forever) regardless of the value of d. However, it is presently beyond the resources of mathematics to prove this. The hypothesis that all such sequences eventually reach one regardless of the value of d is known as the "Collatz Conjecture."

Using a while loop in a script, find out how many iterations it takes for the sequence to reach 1 for $d = 27$ and plot S_n versus n for $d = 27$. Try a few other values of d. Can you find one for which the sequence takes an especially long time to reach 1? What is the longest sequence you can find for starting values d under 1,000?

A.11 Write a function that plots the path of a ray of light through a plano-convex lens of finite focal length f whose flat surface faces upstream (toward the incoming ray). The incident light is parallel to the optic axis at a height h above the axis. h and f should be inputs to the function.

A.12 Write a function that takes a data set consisting of two vectors $x = [x_1, x_2, \ldots, x_N]$ and $y = [y_1, y_2, \ldots, y_N]$, plots points at the coordinates $(x_1, y_1), (x_2, y_2), \ldots, (x_N, y_N)$, and then fits them to a straight line. Display the "best fit" line and put the corresponding best-fit slope in the title of the image. You may not use a preexisting fitting routine. Instead, write your own. *Hint*: One way to get a best-fit line $y_{\text{fit}}(x)$ is to simply try a large number of candidate lines and try to minimize the net vertical distance from the points to the line. That is, minimize: $\sum_{i=1}^{N} (y_i - y_{\text{fit}}(x_i))^2$. You may assume that the fit lines always go through the origin so you only have to vary the slope of the line. Automate this "trial and error" process using a loop.

Appendix B **Computer Code**

The code in this chapter is intended to give you ideas for your own code and a framework upon which to build. I've made an effort to comment the code well so you can see what each line does. I've only listed the Matlab versions of the routines. Both Matlab and Python versions are available at

https://github.com/CambridgeUniversityPress/FirstCourseLaboratoryOptics.

B.1 Image Processing Functions

```matlab
% ----------------------------------------------------------------------------------------
% MAKE AN IMAGE CUT
% Language: Matlab or Octave
% ----------------------------------------------------------------------------------------
% This script loads an image from a file whose full or relative path is specified. The cut
% is made vertically through the center of the image. Then the vector containing
% the cut intensities is smoothed to reduce laser speckle and displayed.
% ----------------------------------------------------------------------------------------


% ---------------------------------------------------------
% Read in the image and convert it to grayscale if necessary
% ---------------------------------------------------------

imagefile = 'myphoto.jpg';          % User specified path to image file
A=imread(imagefile);                % image is read in as an array A
A=double(A)/255;                    % doubles needed for most purposes, also normalize

if length(size(A))==3               % Color array has three pages (for R, G, & B  resp.)
    A = sum(A,3)/3;                 % sum over pages (3rd array dim.) & renormalize.
end

% ---------------------------------------------------------
% Cut out the desired column and smooth the data
% ---------------------------------------------------------
col = round(size(A,2)/2);           % the column index corresponding to the image center
cut = A(:,col);                     % this is how the cut is actually taken

softcut = movmean(cut,10);          % smooth out laser speckle using a moving mean

% ---------------------------------------------------------
% Display the smoothed and unmodified intensity
% ---------------------------------------------------------
plot(cut); hold on;                 % plot the original data
plot(softcut,'linewidth',2);        % and the smoothed version
hold off;
```

```
% ---------------------------------------------------------------------------------
% REMOVE LASER SPECKLE VIA FFT
% Language: Matlab or Octave
% ---------------------------------------------------------------------------------
% Performs a 2D fourier transform on the image in "imagefile", then sets to zero, all spatial
% frequency bins with amplitude below the value given in  "threshold". The result is then
% fourier transformed back to the spatial domain to produce the despeckled image. The beam
% occupies a small part of the freq. domain image; speckle occupies the rest. Speckle has low
% power per frequency bin. Suppressing all bins with low power therefore gets rid of speckle.
%
% SYNTAX:  I = despeckle(imagefile, threshold);
%
% NOTES: - threshold = 1 is usually a good starting point.
%        - The image is assumed to be a monochrome image. If it is not, then it is converted
%          to a monochrom image.
%
% INPUT ARGUMENTS
% ---------------
% imagefile:   Full path and filename to the image (any format readable by "imread").
% threshold:   Frequency components with log10(mag) below threshold are discarded.
% ---------------------------------------------------------------------------------

function despekld_image = imdespeckle(imagefile, threshold)

% The following lines read in the figure data, convert the color image to intensity and
% convert the uint8 data type of the figure to the double (real numbers) data type used
% for computation.
data = imread(imagefile);
if size(data,3)>1, data=rgb2gray(data); end
data = double(data);

% Perform the 2D numerical fourier transform and scale it correctly. The result is a
% picture of the image in "frequency space" (spatial frequency, that is).
N1 = size(data,1); N2=size(data,2);
F = fftshift(fft2(data)/sqrt(N1*N2));

% Threshold the fourier transformed image
pixels_below_threshold = log10(abs(F))<threshold;   % logical mask for pixels -> 0
Fthresh=F;                                           % start unthresholded
Fthresh(pixels_below_threshold)=0;                   % set pixels below threshold to 0

% Finally, perform the inverse transform on the thresholded data to get back
% to position space. (I.e. to get back our image.).
despekld_image = abs(ifft2(Fthresh)*sqrt(N1*N2));

% The plotting is done by the surf command. (There are also numerous figure options
% called using commands like: view, set, axis, colorbar, and caxis. But these are
% only to make the picture prettier, easier to rotate in 3D, etc.)

figure(1);
h1=surf(abs(despekld_image));
set(h1,'linestyle','none');
colormap(bone);
view(2);
set(gca,'color',[1 1 1]*0.3,'plotboxaspectratio',[N2/N1,1,1],...
    'ydir','reverse','xaxislocation','top','fontname','fixedwidth');
axis tight;
axlims = axis;
title('Despeckled Image','fontsize',9,'fontname','fixedwidth');

figure(2);
h2=surf(log10(abs(Fthresh)));  set(h2,'linestyle','none');
view(2);
set(gca,'color',[1 1 1]*0.3,'plotboxaspectratio',[N2/N1,1,1],...
    'ydir','reverse','xaxislocation','top','fontname','fixedwidth');
axis square;
axis(axlims);
colorbar;
title('2D FFT, Thresholded & Smoothed','fontsize',9,'fontname','fixedwidth');
```

B.2 Miscellaneous Functions

```
% --------------------------------------------------------------------
% FIT FUNCTION FOR BEAMWIDTH MEASUREMENTS
% Language: Matlab and Octave
% --------------------------------------------------------------------
% Returns the field radius of a TEM_00 mode beam at any point z
% along the optic axis. Fit parameters are beam width. The input
% arguments w0, zw, lambda, z all need to be in the same units.
% The output arguments will be in those units.
%
% SYNTAX: [w,R,zR]=beamradius([w0,zw,lambda],z);
%
% w0 = waist size
% zw = position of waist
% lambda = wavelength
%
% w  = spot size (field radius) at z
% R  = curvature of phasefront at z
% zR = Raleigh length.
% --------------------------------------------------------------------


function [w,R,zR]=beamradius(params,z)

w0=params(1);                          % beam (field) width at waist [meters]
zw=params(2);                          % waist position [meters]
lambda=params(3);                      % wavelength [meters]
zR=pi*w0^2/lambda;                     % Raleigh length [meters]

w=w0.*sqrt(1+((z-zw)/zR).^2);          % beam width at z [meters]

if nargout>=2
    R=z.*(1+(zR./z).^2);               % beam phasefront curvature at z
end




% ----------------------------------------------------------------
% GET THE Q-FACTOR OF A BEAM
% Language: Matlab and Octave
% ----------------------------------------------------------------
% Returns the q-factor of a Gaussian beam given the spot size, w,
% phasefront radius of curvature, R, and wavelength, lambda.
%
% SYNTAX: qfactor=q_(w,R <,lambda>);
%
% w      = 1/e Field radius
% R      = Radius of curvature of phasefront
% lambda = wavelength
%
% Any one of w, R and lambda may be a vectors or scalars.
% If more than one of w, R and lambda is a vector, all
% vectors supplied must be the same size. w, R and lambda must
% all be in the same units.
% ----------------------------------------------------------------

function qfactor=q_(w,R,varargin)

if nargin>=3, lambda=varargin{1}; else lambda=1064e-9; end

if R~=Inf
    qfactor=pi*w.^2.*R./(pi*w.^2-1i.*R.*lambda);
else
    qfactor=1i*pi*w.^2./lambda;
end
```

```
% ------------------------------------------------------------------
% GET BEAM WIDTH AND ROC FROM Q
% Language: Matlab and Octave
% ------------------------------------------------------------------
% Returns the phasefront radius of curvature and the beam width
% given the q factor of a Gaussian beam and the wavelength.
%
% SYNTAX: [R <,w>]=R_(q <,lambda>);
%             <...> indicates optional arguments
%
% q     = q-factor of the beam at the position where R and w are to
%         be found. q can be a vector
% lambda = wavelength. Can be a vector or scalar.
%
% If both q and lambda are vectors, they must be the same size.
% If w is requested as an output, lambda should be supplied.
% ------------------------------------------------------------------

function [Rout,wout]=R_(q,varargin)

if nargin>=2, lambda=varargin{1}; else lambda=1064e-9; end

wout=sqrt(lambda/pi .* imag(q).*(1+real(q).^2./imag(q).^2));
Rout=real(q).*(1+imag(q).^2./real(q).^2).*ones(size(wout));
```

```
% --------------------------------------------------------------------
% PROPAGATE A GAUSSIAN BEAM
% Language: Matlab and Octave
% --------------------------------------------------------------------
% Propagates a Gaussian beam (TEM_nm) with complex radius of curvature q1
% and amplitude factor p1 (optional), according to the ABCD matrix
% supplied.
%
% Returns the new complex beam radius q=(A*q1+B)/(C*q1+D) and the
% new amplitude factor p = 1/(A+B/q1)^(1+n+m) by which the field is
% multiplied. If q1 is a vector q and p will be vectors of the same size.
%
% SYNTAX: [q,p]=prop(q1,abcd <,[n,m],p1>);
%             <...> indicates optional arguments
%
% For a Hermite Gaussian n,m are the mode designators.
% --------------------------------------------------------------------
function [q,p]=prop(q1,abcd,varargin)

if ( nargin>=3 && ~isempty(varargin{1}) ), mode=varargin{1}; else mode=[0,0]; end
if nargin>=4, p1=varargin{2}; else p1=ones(size(q1)); end

A=abcd(1,1);
B=abcd(1,2);
C=abcd(2,1);
D=abcd(2,2);

n=mode(1);
m=mode(2);

q = (A*q1 + B)./(C*q1 + D);
p = p1.*exp(i*angle(1./(A+B./q1).^(1+n+m)));
```

B.3 Curve Fitting

```
% --------------------------------------------------------------------------------
% DATA FITTING DEMONSTRATION
% Language: Matlab (see Line 158 for modifications needed to run under Octave)
% --------------------------------------------------------------------------------
% This script makes some simulated data for the intensity at the output of a michelson as a
% function of mirror position with added noise.  It then fits the data to a straight line
% using one of Matlab's built-in curve-fitting tools: lsqnonlin. It handles non-uniform
% uncertainties correctly and propagates the data uncertainties into the best fit parameters.
% --------------------------------------------------------------------------------


%------------------------------------------------------------------
% Make the simulated data
%------------------------------------------------------------------

x = linspace(-5,5,9).';             % x is the change in the length of one arm
x = x + 0.2*randn(size(x));         % Make the vector unevenly spaced

a1_actual = 1/2*(1+(4*pi/633)*2);   % Offset (2 nm phase offset)
a2_actual = 2*pi/633;               % Slope (nm^-1). Michelson made with a HeNe
y = a1_actual + a2_actual*x;        % The physical relationship betw. x & y

sigma_y = 0.025;                    % The std of y values (the actual uncertainty)
y = y + sigma_y*randn(size(y));     % Add the fluctuations due to the uncertainty
yerr =  abs(y)*0.02;                % 2% uncertainty estimate made by the observer


%------------------------------------------------------------------
% Display the data
%------------------------------------------------------------------

format compact
format short
disp([x,y,yerr]);                   % The experimenters lab book table
figure(1);                          % Open a figure in which to plot
h=errorbar(x,y,yerr,'d');           % Plot the data with the unc. estimates
set(h,'linewidth',1.5);             % Make the plot lines a bit thicker
set(gca,'xlim',[-10 10]);           % Choose the x axis limits of the plot
set(gca,'ylim',[0.4 0.65]);         % Choose the y axis limits of the plot]
xticks(-10:2:10);                   % Make the x axis have ticks on the even integers
grid on;                            % draw the grid
set(gca,'fontsize',14);             % Bigger default font for this plot
box off                             % No box around the plot
xlabel('Mirror motion (nm)');       % Label the x axis
ylabel('P_{out} / P_{one arm}');    % Label the y axis
hold on;                            % Allow the next plot to share the same axes


%------------------------------------------------------------------
% Fit the data to "fitfunc"        % see end of script for functions
%------------------------------------------------------------------

a0 = [0;1];                         % your initial guess at the best fit values
[a,~,res,~,~,~,jac] = ...           % "a" are the best fit values, res and jac
    lsqnonlin(...                   % are used to find the uncertainty in a(1), a(2), ...
        @(a)weighted_residuals(a,@fitfunc,x,y,yerr),... % this "anonymous function" is minimized
        a0...                       % Note: @(a)weigh... defines the anonymous function of a
        );                          % = weighted_residual for the x,y,yerr data given
xfit = linspace(-8,8,100);          % x values for plotting the fit function
plot(xfit,fitfunc(a,xfit),'--',...  % Plot the fit using the same function to
    'linewidth',2);                 % Generate the y-values as was used in the fit
hold off;                           % Allow the next plot to wipe the current one
legend('Data','Best fit line',...   % Put a legend in the lower right corner
    'location','southeast');        % (a.k.a the "southeast" corner)
title('Data and fit');
X2red = 1/(length(x)-length(a))*sum(...  % The reduced chi-squared of the best fit...
    (y-fitfunc(a,x)).^2./yerr.^2);       % should be close to 1.
```

```
disp(['X2red = ',num2str(X2red)]);              % In the Matlab command window


%----------------------------------------------------------------
% Direct estimate of the uncertainties from X2 curvature
%----------------------------------------------------------------
J=zeros(length(a),1); da=zeros(length(a),1); % Set up the variables
for r=1:length(a)
    Jsqr(r)=sum( (jac(:,r)).^2 );             % curvature of the a_r chi-square cut at the minimum
    da(r) = 1/sqrt(Jsqr(r));                  % is approx C_r =  2*sum(jac(:,r).^2. Gives parabola.
end
disp('Solution +/- uncertaintes:')
disp([a,da]);                                 % display best-fit values and uncertainties


%----------------------------------------------------------------
% OPTIONAL:
%
% Make and display chi-square cuts for each fit variable. Also,
% draw the line corresponding to chi-square increasing by 1.
%----------------------------------------------------------------

a_1 = linspace(0.4,0.6,200);            % The domain of a_1-axis chi-square cut
a_2 = linspace(0.005,0.015,200);        % The domain of a_2-axis chi-square cut
X2_a1cut = zeros(size(a_1));            % will hold the y-values of the a_1 cut
X2_a2cut = zeros(size(a_2));            % will hold the y-values of the a_2 cut
for s = 1:length(a_1)
    X2_a1cut(s) = ChiSqr([a_1(s),a(2)],@fitfunc,x,y,yerr); % the a_1 cut chi-square values
end
for s = 1:length(a_2)
    X2_a2cut(s) = ChiSqr([a(1),a_2(s)],@fitfunc,x,y,yerr); % the a_2 cut chi-square values
end

figure(2);                                      % this figure 2 will hold two plot windows
subplot(2,1,1);                                 % this sets up the first of two plot windows
h=plot(a_1,X2_a1cut,'-',a(1),min(X2_a1cut),'o',...  % plot X^2 cut in the a_1 dir, the min, ...
    [min(a_1),max(a_1)],[min(X2_a1cut),min(X2_a1cut)]+1,'--');  % and min+1 line
set(h,'linewidth',2);                           % use bolder lines
grid on;                                         % draw a grid on the plots
xlabel('a_1','fontname','Times New Roman','fontangle','italic');    % x-axis label
ylabel('\chi^2');                               % y-axis label
set(gca,'fontsize',16);                         % make the font size bigger
title('\chi^2 cuts');                           % add a title to the graph
subplot(2,1,2);                                  % set up the second plot window
h=plot(a_2,X2_a2cut,'-',a(2),min(X2_a2cut),'o',...  % plot X^2 cut in the a_2 dir, the min, ..
    [min(a_2),max(a_2)],[min(X2_a2cut),min(X2_a2cut)]+1,'--'); % and min+1 line
set(h,'linewidth',2);
grid on;
xlabel('a_2','fontname','Times New Roman','fontangle','italic');
ylabel('\chi^2');
set(gca,'fontsize',16);


%----------------------------------------------------------------
% OPTIONAL:
%
% Generate & Display the chi-squared surface
% (This section for illustrative purposes and can be omitted.)
%----------------------------------------------------------------

[a1,a2] = meshgrid(...                          % Set up the domain. a1 and a2 are
    linspace(0.47,0.55,100),...                 % matrixes of coordinates (parameters)
    linspace(-0.01,0.03,100)...                 % at which to calculate chi-squared
    );

X2 = zeros(size(a1));                           % This is the surface we will be finding
for k=1:size(a1,1)                              % step through all the values of a1 and a2
    for s=1:size(a2,2)                          % in the desired range
        X2(k,s) = ChiSqr([a1(k,s),a2(k,s)],@fitfunc,x,y,yerr); % Uses ChiSqr function def. below
    end                                          % Formula for the chi-squared
```

```
end

figure(3);                              % Open a figure to hold the chi-sqr plot
pcolor(a1,a2,X2);                       % plot the chi-square surface.
colormap(gray(20));                     % White is high, black is low
shading interp;                         % Makes it a bit smoother
caxis(min(min(X2))+[0,100])             % Set the color range
cbar=colorbar;                          % Color key
set(gca,'fontsize',16);                 % Bigger fonts are more visible
ylabel(cbar,'   \chi^2','rotation',0,'fontsize',16); % y-axis text label
daspect([1,1,1]);                       % Make the axes equally spaced
hold on;
contour(a1,a2,X2,min(min(X2))+[1,1],'w:','linewidth',2); % 1 contour at min(X2) + 1
plot(a(1),a(2),'w.');                   % Best fit values of a1 and a2
xlabel('a_1','fontname','Times New Roman','fontangle','italic');
ylabel('a_2','fontname','Times New Roman','fontangle','italic');
title('\chi^2 surface');
hold off


%------------------------------------------------------------------
% Functions used in script (these must be at the end of the script
% file in Matlab but at the beginning of the script file in Octave.)
%------------------------------------------------------------------

1;                                      % (Octave only) script can't start with function def.
% The function which we fit to the data
function y=fitfunc(a,x)
y=a(1)+a(2).*x;                         % straight line (can be changed to anything)
end

% This is the quantity whos least square is to be minimized
function r = weighted_residuals(a,fhandle,x,y,yerr)
r=(feval(fhandle,a,x)-y)./yerr;         % gets the fit function via its "handle"
end

% Convenience function for calculating the chi-square
function C = ChiSqr(a,fhandle,x,y,yerr)
wr = weighted_residuals(a,fhandle,x,y,yerr);% chisqr is just the quadrature sum of ...
C = sum(wr.^2);                         % the weighted residuals
end
```

B.4 Fourier Propagation

The following Matlab code propagates the electric field in one plane to another downstream plane using the Fourier transform method.

```
% ------------------------------------------------------------------------------
% FOURIER OPTICS DEMONSTRATION
% Language: Matlab or Octave
% ------------------------------------------------------------------------------
% Calculates the optical intensity on a screen due to aperture diffraction.
% The complex scalar field u(x,y) in the  source plane is a concave sperical
% phasefront (ROC=0.5 m) passing through a circular aperture. The resulting
%complex scalar field amplitude u'(x',y') in the "field plane" is calculated
% via Fourier transform.
% ------------------------------------------------------------------------------


% --------------------
% Physical Parameters
% --------------------
c = 3e8;                                    % speed of light in m/s
epsilon0 = 8.854e-12;                       % vacuum permittivity in F/m
lambda = 633e-9;                            % optical wavelength in m

% -------------
% Source plane
% -------------
xpmax=0.002; ymax=xpmax;                    % Src plane area: 4*xmax*ymax m^2
Nx = 2^nextpow2(512); Ny = 2^nextpow2(Nx); % #pts in source plane grid = Nx*Ny
dxp = 2*xpmax/(Nx-1);  dyp=2*ymax/(Ny-1);  % interpixel dist. in the src plane (m)
xp = repmat( ((0:Nx-1)-floor(Nx/2)) *dxp, Ny,1); % x' values at which to calc. source field
yp = repmat( ((0:Ny-1)-floor(Ny/2)).'*dyp, 1,Nx); % y' values at which to calc. source field

% -----------------------
% ABCD Matrix Components
% -----------------------
% Optical system consists of [ FREE SPACE : LENS : FREE SPACE ].
L1 = 0.035 ;%0.1e-3;                        % aperture-lens dist. in m
L2 = 0.05;                                  % lens-screen dist. in m
f  = -0.03;                                 % f=Inf corresponds to no lens
M = [[1 L2];[0 1]] * [[1 0];[-1/f 1]] * [[1 L1];[0 1]]; % ABCD matrix of the system
AA = M(1,1); BB = M(1,2); CC = M(2,1); DD = M(2,2);     % The components A, B, C, D

% ---------
% Aperture
% ---------
% Field amplitude is non-zero at these values of x, y (i.e. where it passes
% through the aperture). The apertures are defined as logical matrixes that
% are used to index the source field distribution, i.e. Usource(~aperture)=0;
% UIsource(aperture)= <something nonzero>.

% a = 50*1e-6;                              % circular obstrution diam. (m)
% b = 600e-6;
% aperture = (xp+0.75*b).^2+(yp-0.35*b).^2 > (a/2)^2; % circular obstruction logical mask

a = 3000*1e-6;                             % equil.triang. aperture side (m)
aperture = ~((yp<sqrt(3)*xp+a/2/sqrt(3)) &...
            (yp<-sqrt(3)*xp+a/2/sqrt(3)) &...
            (yp>-a/2/sqrt(3)));            % equil. triangular aperture

% a = 300e-6;                              % triangular diam. (m)
% b = 600e-6;
% aperture = ~(((yp-0.35*b)<sqrt(3)*(xp+0.75*b)+a/2/sqrt(3)) &...
%             ((yp-0.35*b)<-sqrt(3)*(xp+0.75*b)+a/2/sqrt(3)) &...
%             ((yp-0.35*b)>-a/2/sqrt(3)));           % equil. triangular obstructin

% -------------
% Source Field
```

```
% ------------
% Here, the incident field is assumed to be a Gaussian beam of width "w"
% and radius of curvature "roc". The beam is clipped by the aperture.
roc = 0.5;                                          % R.O.C. of phasefront at src plane (m)
w  = 750e-6;                                         % beam width of incident beam (m)
I0 = 1e6;                                            % max src plane intensity (W/m^2)
E0 = sqrt(2*I0/c/epsilon0);                          % m ax field ampl. in src plane (N/C)
k=2*pi/lambda;                                       % wave number
r=sqrt(xp.^2+yp.^2);                                 % src plane coordss dist from center
usource = E0*exp(-r.^2/w^2).*exp(1i*k*r.^2/2/roc);   % field ampl. in src plane
usource(~aperture)=0;                                % field is zero except in the aperture
Isource = epsilon0*c/2*abs(usource).^2;              % Intensity in the source plane (W/m^2)


% ==========================================================================
% |+|+|+|+|  THE COMPUTATION OCCURS BETWEEN THIS LINE AND THE ONE LIKE IT BELOW  |+|+|+|+|
% ==========================================================================

% h, below is a scale factor to change from the physical units (meters) to new units in
% which all physical lengths are scaled by h=sqrt(B*lambda). In the new units, the Fresnel
% integral becomes a standard Fourier transform multiplied by a phase factor. We now scale
% all physical lengths to the new units before performing the fourier tranform. Due to the
% limitations placed on variable names, x' in the text is the variable f here, y' is g,
% X' is F, and Y' is G.

h = sqrt(BB*lambda);                                 % scaling factor
dXp=dxp/h; dYp=dyp/h;                                % src interpixel dist in the new units
Xp = xp/h;                                           % src plane x-coords scaled to new units
Yp = yp/h;                                           % src plane y-coords scaled to new units

dX = 1/dXp/Nx;  dY=1/dYp/Ny;                         % corresponding spatial sampling interval
                                                     % in field plane after 2 dim. FFT (fft2).
X=repmat(((([0:Nx-1]-floor(Nx/2))  *dX,Ny,1);        % Field plane, x-domain (in scaled length)
Y=repmat(((([0:Ny-1]-floor(Ny/2)).'*dY,1,Nx);        % Field plane, y-domain (in scaled length)
dx=dX*h;  dy=dY*h;                                    % field plane sampling interval (in meters)
x = X*h;  y = Y*h;                                    % Field plane, x and y-domains (in meters)

% Perform 2D FFT on and scale correctly
% ------------------------------------
ufield = ...
    -1i*exp(1i*pi*DD/BB/lambda*((x).^2+(y).^2))...   % Perform the 2D FFT on the field
    .*fftshift( fft2( exp(1i*pi*AA*(Xp.^2+Yp.^2)).*usource )*dXp*dYp ); % FT2
Ifield = epsilon0*c/2*abs(ufield).^2;                % get the intensity


% ==========================================================================
% |+|+|+|+|  CODE BELOW CHECKS AND DISPLAYS THE RESULTS |+|+|+|+|+|+|+|+|+|+|+|+|+|+|+|+|
% ==========================================================================

% Check energy conservation
% -------------------------
inpow =  trapz(trapz(Isource))*dxp*dxp; % integral of intensity in the src plane
outpow = trapz(trapz(Ifield))*dx*dy; % (total power) should equal field plane
disp(['Power in the source plane:  Pin  = ',num2str(inpow*1000),' mW']);
disp(['Power in the field plane:  Pout = ',num2str(outpow*1000),' mW']);

% Make a red colormap for use in displaying the laser beam intesity
% ----------------------------------------------------------------
cb=sqrt(colormap('bone'));                           % modify the built-in 'bone' colormap
cmred=[cb(:,2)*1, cb(:,2)*0.2, cb(:,2)*0.1];         % make it shades of red rather than grey

% Display source plane intensity (Fig. 1)
% ---------------------------------------
figure(1);                                           % open a figure window
ax1 = surf(xp*1e3,yp*1e3,(Isource/1000));            % intens. (mW/mm^2) in src pl. (x,y in mm)
view(2);                                             % top-view orientation
xlabel('x (mm)');                                    % label the axes
ylabel('y (mm)');
axis square                                          % show both axes on the same scale
axis tight                                           % minimize white space
set(ax1,'linestyle','none');                         % don't draw the axes
caxis([min([max(Isource(aperture)),...               % set the color axis
```

```
        min(Isource(aperture))]),max(Isource(aperture))]/1000);
colormap(cmred);  % activate the color map
shading interp; % looks more realistic
grid off;
cbar1=colorbar;
ylabel(cbar1,'Intensity (mW/mm^2)');
hold off;
title('Source Plane Intensity');
set(gca,'fontsize',14);

% Display field plane intensity (Fig. 2)
% ------------------------------------
figure(2);
ax2 = surf(x*1e3,y*1e3,(Ifield/1000));          % plot the intensity in the field plane
view(2);                                         % can be rotated from this top-view
shading interp;
grid off;
xlabel('x (mm)');
ylabel('y (mm)');
axis square
axis tight
set(ax2,'linestyle','none');
caxis(([min(min(Ifield)) (max(max(Ifield)))/1000]));
title('Diffracted Intensity in the Field Plane');
colormap(cmred);
cbar2=colorbar;
ylabel(cbar2,'Intensity (mW/mm^2)');
set(gca,'fontsize',14);
```

References

52 teams of gravitational wave, electromagnetic, and neutrino astronomers, Abbott, B. P., et al. 2017. Multi-messenger observations of a binary neutron star merger. *The Astrophysical Journal Letters*, **848**(2), L12.

Bachor, Hans-A. and Ralph, Timothy C. 2019. *A guide to experiments in quantum optics*. Weinheim: Wiley-VCH.

Bass, Michael. 2010. *Handbook of optics, third edition, Vol. 1*. New York: McGraw-Hill.

Bennett, Charles A. 2008. *Principles of physical optics*. Hoboken, NJ: Wiley.

Bevington, Philip. 2003. *Data reduction and error analysis for the physical sciences*. Boston, MA: McGraw-Hill.

Beyersdorf, Peter 2014. *Laboratory optics: A practical guide to working in an optics lab*. Google eBook.

Born, Max and Wolf, Emil. 2019. *Principles of optics*. Cambridge: Cambridge University Press.

Fowles, Grant. 1989. *Introduction to modern optics*. New York, NY: Dover Publications.

Griffiths, David. 2017. *Introduction to electrodynamics*. Cambridge and New York: Cambridge University Press.

Hecht, Eugene. 2017. *Optics*. Boston, MA: Pearson Education, Inc.

Horowitz, Paul. 2015. *The art of electronics*. New York: Cambridge University Press.

James, J. F. 2014. *An introduction to practical laboratory optics*. Cambridge and New York, NY: Cambridge University Press.

LIGO and Virgo Scientific Collaborations, Abbott, B. P., et al. 2016. Observation of gravitational waves from a binary black hole merger. *Physical Review Letters*, **116** (February), 061102.

Magaña-Sandoval, Fabian, Vo, Thomas, Vander-Hyde, Daniel, Sanders, J. R., and Ballmer, Stefan W. 2019. Sensing optical cavity mismatch with a mode-converter and quadrant photodiode. *Physical Review D*, **100**(10), 102001.

Michelson, Albert Abraham. 1903. *Light waves and their uses*. Chicago, IL: The University of Chicago Press. Chap. Lecture IV The application of interference methods to spectroscopy.

Pedrotti, Frank. 2007. *Introduction to optics*. Upper Saddle River, NJ: Pearson Prentice Hall.

Saulson, Peter R. 2017. *Interferometric gravitational wave detectors*. 2nd ed. Toh Tuck Link, Singapore: World Scientific.

Siegman, Anthony E. 1986. *Lasers*. Mill Valley, CA: University Science Books.

Svelto, Orazio 2010. *Principles of lasers*. New York: Springer.

Yariv, Amnon. 2007. *Photonics: Optical electronics in modern communications*. New York and Oxford: Oxford University Press.

Index

Printed in the United States
by Baker & Taylor Publisher Services